DATE DUE

NO 18 '97			
DE 1 '97			
DE 19 '97			
MY 19'00			
AG 10			
OC 24 05			
DE 16 05			

DEMCO 38-296

Conservation and Biodiversity

Andrew P. Dobson

SCIENTIFIC
AMERICAN
LIBRARY

A division of HPHLP
New York

Library of Congress Cataloging-in-Publication Data

Dobson, Andrew P.
    Conservation and biodiversity   /   Andrew P. Dobson.
        p.   cm.
    Includes bibliographical references and index.
    ISBN 0-7167-5057-0 (hard cover)
    1. Biological diversity conservation.   I. Title.
    QH75.D59   1995
    333.95′16—dc20   95-24711
                                                                          CIP

ISSN   1040-3213

Printed in the United States of America

Scientific American Library
A division of HPHLP
New York

Distributed by W. H. Freeman and Company,
41 Madison Avenue, New York, New York 10010
Houndmills, Basingstoke, RG21 6XS, England

First printing 1995, HAW

This book is number 59 of a series.

Contents

Preface

Biologists find themselves in a strange position at the end of the twentieth century. Never has a real understanding of our subject seemed so within our reach. A host of new molecular biological techniques are allowing us to address in great detail the mechanisms by which life evolved on earth, while the speed and ease of modern travel make it possible for us to study almost any species or ecological system we wish. Yet excitement at this scientific progress is tempered by realization that much of earth's life is endangered. At no other time in the world's history have species and natural habitats been destroyed at such a rapid rate.

The grapevine of conservation biology seems to always carry some new report of another vanished species: a frog in Costa Rica, a fish in the Great Lakes. One of the challenges faced by conservation biologists is to transform rumor into rigorous, defensible estimates of extinctions. How can we determine whether rates of extinction are actually increasing, and, if they are, can we determine by how much? Can we quantify how the loss of forests and other natural habitats are causing reductions in the diversity and abundance of other species on the planet? Only by addressing these questions can we gauge the urgency of conservation measures and develop the optimal strategies for achieving conservation goals.

These questions lead to deeper investigations into the magnitude and value of the earth's biodiversity, which can be understood as the total variety of life on earth, including not only the many plant, animal, fungal, and bacterial species, but also the large amount of varia-tion that exists within each species at the genetic and individual levels. In *Conservation and Biodiversity*, I attempt to provide an introduction to the problems, both scientific and economic, of quantifying the magnitude and value of biodiversity, as well as answer questions about the current rates of loss of biodiversity.

If species are disappearing as quickly as it seems, what do we do about it? An important aim of this book is to describe the scientific principles underlying techniques for saving what remains. Conservation biologists are approaching the problem in a variety of ways. Using satellite imagery, we are able to locate and map potential sites for nature reserves with increasing accuracy; as land is set aside for reserves, we are finding new ways to monitor the populations and communities of species that live inside reserves while minimizing disruption of their normal day-to-day activities. Mathematical biologists are wrestling with the problems of population dynamics and discovering intellectual challenges that exercise their imaginations in ways not seen even in the rocket sciences. The world's zoos and botanical gardens are developing into self-sufficient conservation centers, which no longer remove species from the wild, but instead play a major role in educating people, children particularly, about the natural diversity of the world in which we live.

The achievements of those who helped initiate conservation biology as a science are an inspiration to an increasing number of younger scientists and activists who seek to emulate them. The conservation of the earth's biodiversity is turning out to be a greater and more complex scientific challenge for them than the Manhattan project was for their grandparents. Their goal is not only to ensure that biodiversity is conserved but to do so in a way that respects human cultural values. The exponential growth of the Society for Conservation Biology is a hopeful sign for the future of conservation.

T. R. E. Southwoods's lectures to me as a first-year undergraduate at Imperial College first stimulated my interest in the environment. The ideas and examples presented in this book were then thought about, discussed, and written down in a variety of locations: with Chas MacLellan and Pete Wallen in a number of pubs in Yorkshire (The Three Tuns, Thirsk), Oxford (The Eight Bells, Cumnor, and The Rose and Crown), and London (The Churchill); at Valera Lyles house in Carmel, California; Iain and Oria Douglas-Hamilton's farm at Lake Naivasha; Joyce and Selengai Poole's house at Karen and in the Ngongs, Kenya; Cynthia Moss's camp at Amboseli, Kenya; Pete and Mary Hudson's house in Newtonmore, Scotland; "Baxter's" in Bozeman, Montana; and at Bryan Grenfell's and Andrew Sugden's houses and the Cambridge Blue in Cambridge, England. My mother, Patricia Loe, brother Stewart, sister Julia, and their families were always there for me when I whizzed through England; they all provided the initial environment for the development of the ideas that became this book. I thank all these friends and relations for their wonderful hospitality and many stimulating conversations.

Many of the book's ideas were formed, gestated, and rearranged in the corridors and offices of Eno Hall at Princeton University. Several chapters formed the basis of my lectures in the course "Managing the Global Environment." My colleagues on this course, Steve Hubbell and Alison Jolly, as well as the students and guest speakers, all helped shape my perspective on what can be done to conserve biodiversity. In the fall of 1992 I taught a graduate course on conservation biology during which the students helped me by discussing rough drafts of these chapters. In addition to these students, I would like to thank the following people for insightful and helpful discussions on a variety of occasions: Robin Absher, Jorge Ahumada, Kristin Ardlie, Dorene Bolze, John Bonner, Robbie Brett, Tormod Burkey, Robin Carper, Mick Crawley, Lisa Curran, Giulio Deleo, Saba and Dudu Douglas-Hamilton, Jessica Eberhard, Ivan, Pam, and Sue Hattingh, George Hurtt, Simon Emms, Leila Fishman, Charles Foley, Joshua Ginsberg, Nick Georgiadis, Martha Groom, Ted Gullison, Henry Horn, Peter Hudson, Martha Hurley, Lucia Jacobs, Alison Jolly, Paula Kahumbo, Margaret Kinnaird, Fred Koontz, Wes Krause, Margarita Lampo, John Lawton, Richard Leakey, Simon Levin, Bill and Will Lyles, Liz Losos, Georgina Mace, Mary Meagher, Adina Merenlender, Paul Moorcroft, Cynthia Moss, Philip Muruthi, Bill Newmark, Diana O'Brien, Tim O'Brien, Steve Pacala, Joyce Poole, Ray Rasker, Mellie Reuling, John Robinson, Jean-Paul Rodriquez, Dan Rubenstein, Michael Scott, Andrew Sugden, Julie Sutherland, David Smith, John Terborgh, Gioia Theler, Barbara Tyack, and Pete Wallen. I am particularly grateful to Robert May and David Wilcove for their wide-ranging comments on earlier ideas and outlines.

The penultimate draft was considerably improved after detailed comments from Walter Reid and Dan Simberloff. At the Scientific American Library, Jonathan Cobb and Susan Moran were patient and perceptive editors, and I learned a lot working with them. Travis Amos found some super photographs. Georgia Lee Hadler shepherded the book through the stages of proof, Blake Logan created its interior design, and Ellen Cash acted as production coordinator. For all of this assistance and advice, I am very grateful.

Most of the text was written, rewritten, and discussed in Annarie Lyles's apartment in the Bronx and in our house in Princeton; I can honestly say that any sections that were not influenced by her, or written in her presence, were not worth keeping and were eventually left out of the final manuscript. Walks around the Bronx zoo kept me sane and helped focus my concentration on why it is crucial to conserve the world's biodiversity.

Walking around the neighborhood and chatting to people in Joe's Deli, the Arthur Avenue food market, and Rambaruzzi's coffee shop continued to remind me that human diversity is as important as plant and animal diversity.

* * *

Alain Durning of the WorldWatch Institute has demonstrated the scale of ecological change in an imaginary 10-minute film of the last 10,000 years of the earth's natural history. For the first seven minutes of the film, the earth looks unchanged; it is a blue planet with 34 percent of its land area covered in trees. After seven and a half minutes the lands around the Aegean sea lose their covering of trees; this is the flowering of ancient Greek civilization. This first erosion slowly spreads throughout southern Europe until at nine minutes (one thousand years ago) parts of Europe, China, India, and Central America begin to look a little threadbare. The amoebalike patches of development continue to spread, so that 12 seconds from the end (200 years ago) large parts of Europe, China, and the Caribbean islands are bare; ironically Central America has exhibited a short-term recovery following the collapse of the Maya civilization, 30 seconds previously. Six seconds later, the forests disappear from eastern North America, yet forests still cover 32 percent of the earth's land surface. In the last three seconds (the last 50 years) vast areas of forest disappear from Japan, the Philippines, Southeast Asia, central Europe, Central America, eastern South America, western and southern North America, and the Indian subcontinent. In the final seconds, vast fires burn in the eastern Amazon and forest vanishes from northern Canada and Siberia. At the end of the film only 12 percent of the earth's land surface is covered with pristine forest. More than half the forests, holding more than 50 percent of the earth's species, have been destroyed on a timescale that is approximately equal to one human lifespan. A large number of people will have gained employment from this loss, but there will soon be no forests left for them to cut. A handful of people have been made very wealthy; most of us will be considerably impoverished.

As far as we know, life is the one feature that makes our planet unique, and life's vast diversity is perhaps its most impressive trait. Most of the world's political, economic, and health problems, present and future, are intimately linked to the way we manage the world's immense variety of wildlife and natural resources. It is not too late to save a large amount of the earth's remaining biodiversity, but time is running very short.

Andrew P. Dobson
October 1995

*P*rologue

I magine yourself standing in a vast field of wheat in the Midwestern United States: the landscape is completely dominated by a single genetic strain of a modified grass species stretching to the horizon in every direction. Now imagine yourself sitting on the floor of the Peruvian rainforest (mind the ants!). A large variety of tree species surrounds you; and birds of 20 to 40 species are singing all around, although it's hard to catch more than a glimpse of any one of them. Furthermore, it is almost impossible to find two individuals of the same species (apart from the ants). The more time you spend looking for two identical individuals, the more species you keep finding. These two vistas represent two extreme points on a spectrum of biodiversity, which runs from the rich complexity of the Amazon rainforest to the artificial monotony of an agricultural landscape.

1

▶ A diversity of plant species fills even a small plot of tropical forest, as in this view of the Bogar Botanical Gardens in Java.

It has taken human beings several million years to make the evolutionary journey from forest primates to cereal growers and urban administrators. From many perspectives we have been spectacularly successful in our colonization of the earth and exploitation of its natural resources. Yet from another perspective much remains to be learned about our planet; we are only beginning to understand the magnitude and value of the diversity of its life. Unfortunately, we are arriving at this understanding at a time when the earth is losing species at a rate that is many times faster than at any previous time in its history. Estimates suggest that at least 10 percent of species now living, and perhaps as many as 20 percent, will be driven to extinction in the next twenty to fifty years. We ourselves are the primary cause of these extinctions through our modification of the natural environment to create new lands for agriculture, homes, and industry demanded by a rapidly growing population.

The long-term value of the lost plant and animal species is difficult to estimate, and easy to disregard when we need to plant crops that provide the next year's food supply. Agricultural land provides a tangible crop whose value is readily determined by market forces. In contrast, the earth's remaining natural biological diversity is heterogeneously distributed, and nowhere easy to classify in either taxonomic or economic terms. Farmland is obviously necessary to provide the food we eat, but what do we gain from the diversity of life around us?

Biodiversity provides us, first of all, with a wealth of services, which ensure that the air we breathe is clean and the water we drink is potable. The world's forests, oceans, and wilderness areas absorb the by-products of human agricultural and industrial activity, removing a large portion from the atmosphere and slowing the buildup of carbon dioxide and other greenhouse gases that are steadily causing global climate change. Until recently, the presence of these natural habitats ensured that the earth's climate was maintained in the long-term stable state that has allowed human life to evolve and prosper. If we continue to destroy forests and disrupt the oceans, most scientists believe that this capacity to buffer the earth's climate will be considerably diminished. Moreover, a vast number of other species recycle the gases we breathe, the water we drink, and the nutrients in the food we eat. Ultimately, we are much more dependent upon the existence of other forms of biodiversity than it is upon us.

Paradoxically, as our dominion over natural diversity increases, so does our vulnerability. One way to see how is to compare the resilience of our wheat field and Peruvian forest in the face of a lengthy drought or an outbreak of infectious disease. The genetic uniformity of the wheat field makes it susceptible to devastating epidemics brought on by fungal or viral pathogens. Similarly, in a sufficiently prolonged drought the wheat will wither and die before it produces the year's harvest. In contrast, the pathogens in the rainforest will have difficulty finding hosts of the same species. They can't afford to kill any individual host too rapidly if they wish to continue their life cycle by transmitting infections to new hosts. In the face of drought some forest plants will wither and their populations will decline, but other species will take advantage of the extra light and nutrient resources freed up by the disappearance of the susceptible species.

All of us derive considerably more benefits from the presence of worms and fungi in the soil than we do from any celestial bodies other than the sun. The other stars and planets in the universe provide inspiration for poets, and occasional navigational assistance, but as far as we know they are all unoccupied. Although many people still superstitiously believe that the stars play a greater role in determining our future than worms, fungi, and insects, they are very much mistaken. Yet it is a peculiar aspect of human intelligence that we know more about the structure of the universe than we do about the diversity of worms, fungi, and insects. Perhaps we shouldn't be surprised. In a number of respects, biodiversity, and its conservation, is a more difficult intellectual challenge than many other major scientific problems. The interactions between the hundreds of species that allow an ecosystem to function are complex and do not follow predictable patterns that can be easily described by simple mathematics. At present we are losing species at rates at least a thousand times faster, perhaps ten thousand times faster, than they are being replaced, and yet we are only beginning to

▲ A single plant species dominates this view of a wheat field at harvest time.

understand how simple ecosystems will respond to the loss, or addition, of a single plant or animal species.

Conservation biologists are developing a better scientific understanding of how natural populations and communities of plants and animals interact to form ecosystems that provide the major sources of air, water, and vital nutrients for life on earth, and how these ecosystems behave when disrupted. This ecological understanding must inform a political policy that acknowledges that the quality of human life on earth is directly related to the abundance of biodiversity with which we share the planet. The biggest threat to the planet's remaining biodiversity is human population growth and the concomitant habitat conversion that destroys forest, wetlands, and savannas to create new agricultural lands. The other major threat is our increasing exploitation of natural resources. The competitive market forces that have generated innovation in the development of resources over the last two hundred years have led to rates of exploitation that are between 10,000 to a million times faster than the rate at which these natural resources are renewed. Humans have drawn down the stocks of natural resources such as soil, timber and fisheries, threatening biodiversity directly, and in depleting resources such as coal and oil we produce an indirect threat to biodiversity, since many species may not survive the consequent global raise in temperatures or the release of toxic pollutants.

This book aims to provide a short introduction to the types of scientific problem for which solutions are essential if we are to avoid a mass extinction. The conservation biologist's first task is to quantify biodiversity: to find out the distribution and numbers of species on earth, and the rates at which species are disappearing, and to understand how the loss of forest and other native habitat affects the sizes of the living populations that depend on them.

Conservation biologists can then proceed to develop approaches to saving biodiversity based on scientific principles. A recurring theme of the book will be the interaction between ecology and the economics of sustained use. One goal of conservation biology is the achievement of sustainable use, which has been defined as the use of natural resources that aims to satisfy the needs of present generations without compromising the needs of future generations. To understand how we may exploit the resources offered by nature without damaging ecosystems or endangering an exploited species's survival requires a sophisticated understanding of population dynamics and the use of elaborate computer models. Similarly, the managers of zoos rely on a profound knowledge of genetics and the latest genetics techniques to direct the captive breeding programs that are helping to preserve certain endangered species, while ecologists are developing an extensive knowledge of the complicated interactions among species in order to assist the managers of nature reserves in maintaining the integrity of their systems. The behavior of ecosystems and populations is a fascinating subject of study in its own right; that its study is also vital to the earth's future gives the research of ecologists and conservation biologists an additional urgency.

Chapter 1

What Is Biodiversity?

"Dog!" "Cat!" "Watermelon!" "Dinosaur!" A small child's first excited recognition of other species delights parents as an early indicator of later mental ability. It is soon followed by the child's first attempts to count objects: "One nose, two eyes, lots of fingers. . . ." Aristotle was probably the first scientist to recognize that classifying and counting things is a key to understanding our place in the world. Most of us slowly lose this initial excitement in naming and counting other species; we transfer our enthusiasm to collections of stocks, social conquests, or obscure bits of information about fellow humans. But some people retain and develop their fascination with recognizing other species: birdwatchers keep lists of species they've seen, and some attempt to see all the bird species in a country, or even a continent. My wonderfully eccentric wife keeps a "fish list," partly to confuse ornithologists, but mainly because she likes fish. Although no entomologist could realistically try to see all the insects in a country, one might try to see all the insects in some particular taxonomic group.

◀ Although once common in a variety of habitats in southern Madagascar, ring-tailed lemurs *(Lemur catta)* are now restricted to only a few isolated patches of gallery forest and several large nature reserves.

7

The Swedish biologist Carl Linnaeus made the first systematic list of all known living species in the mid-1700s. A species, the smallest unit of classification commonly used by biologists, can be defined as a group of individuals that are potentially capable of breeding with each other, but not with individuals of other groups (or species). The 1758 edition of Linnaeus's work recorded some 9000 species of animals and plants. There were some problems with his list; for example, misled by their very different plumage, he classified male and female mallard ducks as different species! Nevertheless, Linnaeus's 9000 species provided an important starting point for a list that expanded to over a million species in the next two hundred years. Unfortunately, we still have at least as many new species to classify, and perhaps 10 times as many.

How Do New Species Evolve?

From the fossil record, we know that new species or groups of related species are steadily emerging over geological time (while others are becoming extinct). Ultimately the evolution of new species is the source of the world's biodiversity. To understand the mechanisms that drive the emergence of new species, and create the biodiversity around us, let us consider one component of biodiversity, on one island.

If we were to visit Madagascar, we would find that it contains around 30 species of primates, collectively called the lemurs, that are found nowhere else. These 30 species of lemur are actually 15 fewer than the number that lived there 2000 years ago when humans colonized the island. Although fossil lemurs have been found in a wide variety of regions, including the British Isles and the United States, lemurs are found today only in Madagascar, where they evolved from one or two ancestral lemur species that colonized the island more than 40 million years ago. Indeed, most of the species found in Madagascar today, whether lemurs or others, are *endemic* to the island—that is, they evolved there and are found nowhere else in the world.

Some of these endemic species are descendants of species that were present on Madagascar when the island split off from mainland Africa over 100 million years ago; others, such as the lemurs, are the descendants of small groups that colonized the island at a later stage. Once these populations became established on Madagascar, they would begin to diverge from their ancestral populations on the African mainland. Kept apart by hundreds of miles of water, the island and mainland populations would no longer interbreed. The forces of natural selection would begin to operate on the small genetic differences between individuals in each population, to produce physiological and reproductive differences between the African

▲ Verreaux's Sifaka *(Propithecus verreauxi verreauxi)*, a diurnal lemur species, lives in the spiny forests and gallery forests of south and west Madagascar.

and Malagasy populations. Traits that might have been rare at first, but that allowed each group to utilize more efficiently the food resources available to it, would become commonplace. Eventually the mainland and island populations would become incompatible as breeders, and we could say that a new species was born.

This same process proceeded at a smaller geographical scale within the island. During periods of cooling or warming, the forests shrank, and patches of forest became isolated on different parts of the island. Populations of the one or two ancestral lemur species, separated from one another in far-apart forest patches, diverged and produced more than 50 species of lemur. In contrast, on mainland Africa, the lemur's ancestors were displaced by more competitive and aggressive diurnal monkeys, which themselves diversified and evolved into the modern-day monkeys and apes (and along the way eventually included our own ancestors).

Geographical isolation encouraged the other taxonomic groups in Madagascar to diversify as well, so that today 106 of the 250 bird species are endemic, 233 of the 245 reptile species, 142 of the 144 frog species, 110 of the 112 palm species, and around 80 percent of the estimated 8000 angiosperm (flowering) plant species. These 8000 angiosperms represent 25 percent of all the angiosperms in Africa. Indeed, Madagascar has more orchid species than the entire African mainland. In many ways, the island represents a long-term evolutionary experiment, in which a piece of Africa was broken off and placed in partial isolation more than 100 million years ago, then allowed to develop in a different direction. The many millions of years of isolation gave natural selection the opportunity to produce a flora and fauna that is distinct from that found anywhere else in the world.

The perpetual interaction between ecological, evolutionary, geological, and climatological forces is the major mechanism through which life on earth has achieved its ever-changing variety and abundance. These forces are taken to an extreme in isolated groups of oceanic islands such as Hawaii. Here levels of endemism are very high. As many as 1765 species of vascular plants evolved in the Hawaiian islands and were still present when humans first arrived. Between 94 and 98 percent of these plants were found nowhere else on earth; some were produced through the islands' geographical isolation and others when fragments of forests, savannas, and other ecosystems were isolated for long periods of time. We can find a classic botanical example inside the crater of Haleakala volcano on the Hawaiian island of Maui. This location is the only place in the world that we find *Argyroxiphium macrocephalum*, the Haleakala silversword. This species is endemic to Maui and indeed is found only in the crater of Haleakala. Nevertheless, with a total population of around 50,000 individuals, the plant is quite common within this one site. Paradoxically, it is possible for a species to be common in one place, but extremely rare or absent in all other locations.

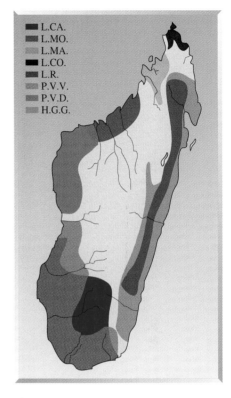

▲ Some lemur species are relatively common and found in several regions of Madagascar, while others are rare and restricted in their range. For example, the spiny forest that characterizes southern Madagascar is home to five lemur species (the ranges of two appear on the map), with the ring-tailed lemur *(Lemur catta)* found throughout this type of habitat. In the tropical rainforest of the eastern part of the island, there may be as many as 14 or 15 lemur species, including the rarely seen aye-aye *(Daubentania madagascensis)*. L. CA.: *Lemur catta*; L. MO.: *L. mongoz*; L. MA.: *L. macaco*; L. CO.: *L. coronatus*; L.R.: *L. rubriventer*; P.V.V.: *Propithecus verreauxi verreauxi*; P.V. D.: *P. verreauxi deckeni*; H. G. G.: *Hapalemur griseus griseus*.

▲ The Haleakala silversword is adapted to survive in the extreme hot and cold temperatures, and dry conditions, of the Haleakala crater, where a warm day may be followed by a night of freezing temperatures, and annual rainfall can be as little as 12 inches. The plant's silvery hairs reflect the sun, and its inward-curling leaves form a bowl that retains rainwater. The tall flower stalk appears just once, and then the plant dies.

What Is Biodiversity?

If we are going to protect biodiversity today, we need to define what we mean by the term and then gain some estimate of its variety by asking how many other species share the planet with us, while also acknowledging that the species composition of the earth is slowly but inexorably changing. The U.S. Office of Technology Assessment has defined biodiversity as "the variety and variability among living organisms and the ecological complexes in which they occur." Alternatively, biodiversity can be defined, in the simplest and most general way, as the sum of all the different kinds of organisms inhabiting a region such as the entire earth, the African continent, the Amazon basin, or our own backyards. At each of these spatial scales the charismatic bird, mammal, and higher plant species—which are relatively easy to observe, photograph, and count—are complemented by a less apparent multitude of bacterial, protozoan, invertebrate, and fungal species that recycle nutrients, clean and rejuvenate air and water, and break down and recycle the waste products and decaying bodies of the larger species.

In its simplest form, quantifying biodiversity requires asking the questions a small child would ask if lost in the forest at night: "Who's there? Where are you? What are you doing?" Our best hope of understanding its magnitude is to break the problem into tractable parts, starting with a small area, such as our own backyard, and then gradually increasing the geographical scale at which we pose these three deceptively simple questions. As conservation biologists we can attempt to quantify the variety of life in the immediate vicinity of the town where we live, on a continent, or on the entire earth. We then have to decide exactly what it is that we count.

Biodiversity can be understood as an assemblage of several hierarchical components: we can count the number of ecosystems, ecological communities, species, populations, or genes in any defined area. Although the conservation and classification of biodiversity is undertaken on all five levels, a sensible first step is to define the smallest fundamental units of its calculus. It could be argued that individual genes (sequences of DNA on chromosomes that code for a specific function) are the fundamental currency of biological diversity. After all, species are ultimately defined by differences in their genes. Moreover, all species require a diversity of genes spread among the population if they are to retain their ability to adapt to changing environments. A gene that is rare, though present, in one environment may be just what is called for if a population is thrust into a new environment.

There are, however, problems with using genes as the units of diversity. It has so far proved possible to make a full accounting of the genes for only a single species of bacteria. Furthermore, the majority of genes are completely unable to function in isolation. Consequently, it seems sensible to use either individual populations or species as the basic units of biodiversity. Separate populations of a species differ slightly in behavior and appearance; in a rapidly changing world, one population may hold the key traits that allows the species to persist or to evolve into a new, more "fit" species. While individual populations receive legal protection in some countries as distinct evolutionary units, it is easier to outline the problems of quantifying biodiversity at the species level.

Unfortunately, most species are not equally common, nor are they evenly distributed across their geographic range. Thus it is very hard to count the total number of species in a country by visiting just one or two locations. Instead we need to survey its fauna and flora regularly at a large variety of sites. We will then be able to see how their abundance changes geographically over the region's expanse and whether the functions that different species perform—such as pollination, nutrient recycling, or pest control—are also undertaken by other species in the region. Certain species, such as tropical forest trees or large marine corals, are themselves the habitats in which other species can live. Those species that provide a living space for other species, or that perform unique functions, are essential in defining a particular type of ecosystem. In contrast, a species whose functions are duplicated by others may be less essential, since if it

◄ Many bird species in this collection in the bird room at the National Museum of Natural History in Washington, D.C. are now extinct in the wild. Our only knowledge of their ecology is the location of the site where the specimens were found.

▶ The relative abundance of butterfly and moth species captured at a light trap at Rothamsted, England, in 1935. Of the 6814 individuals caught, half were from only 6 common species, out of 197 species represented. Not all the common species are shown.

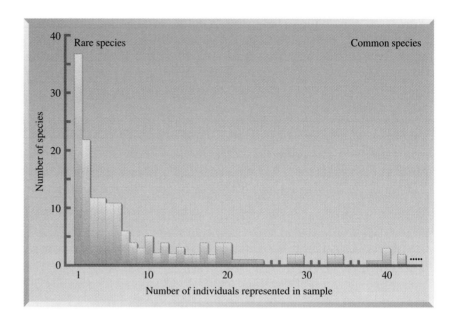

disappears its role may readily be adopted by another species. By developing an understanding of the natural patterns of distribution and abundance of animal and plant species, and the processes that produce these distributions, we take a first crucial step in determining which components of biodiversity can be preserved, and which *must* be saved.

Only a few species, like the house sparrow, are likely to be encountered almost everywhere we turn (and we'll soon tire of counting them). Most species are rare and found only in a few places. Indeed, most species described by taxonomists have been collected from only a single geographical location. Thus, counting the number of species present in any region is a significantly challenging task. Luckily, there are a number of potential shortcuts. We can look for underlying patterns in distribution and abundance that may be common to ecological communities in different regions, at different spatial scales. Using our knowledge of these patterns, we may be able to arrive at estimates of biodiversity from data that are only partially complete.

Any pattern that we observe in nature can be considered a brief snapshot in a long evolutionary history, molded by a variety of evolutionary processes that have continued since life appeared on the planet. There is a subtle interplay between the short-term ecological effects that determine the birth, death, and dispersal of the individuals of a population and the longer-term evolutionary processes that allow some individuals to capitalize upon some small advantage and produce more offspring than others. Life-and-death struggles between individuals of the same, or different, species are the evolutionary engine that drives the spread of novel genetic traits, eventually leading to the production of new species and hence more biodiversity. Ultimately, these interactions determine the continued functioning of the ecosystems upon which all life on earth depends.

Patterns of Abundance at a Single Location

If we were to go out and sample any biological community, we would find, after classifying and counting the numbers of organisms in the community, a remarkable consistency in the structure of the data we collect. The bar graph on the facing page illustrates an example for a community of butterflies and moths sampled one evening at a light trap in an English meadow. The community turns out to be dominated by a few common species, but contains many rare species. Indeed, of the 197 species recorded in the catch, 37 of them are represented by a single individual, while one species forms a quarter of the total.

Ecologists have developed a number of ways of analyzing this remarkably ubiquitous pattern. It is possible, for example, to rearrange the data gathered in the English meadow to create a different form of bar graph. In the graph below, the bars—rather than representing the number of species at each sample size—represent the number of species at each doubling of

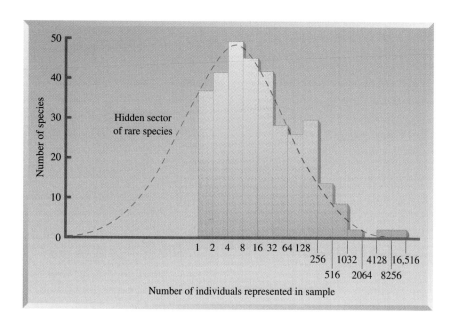

◀ Here the relative abundance of butterfly and moth species captured at a light trap is arranged on a scale of logarithms to the base 2. Thus each division of the bottom axis represents a doubling of abundance, so that the most abundant insect was represented by between 8256 and 16,516 individuals. The curve is truncated at the point where species are represented by a single individual caught at the light trap.

abundance. The tops of the bars trace out a bell-shaped curve, except that the data for the rarest individuals in the community appear to be missing. If a complete bell-shaped curve were to describe the total distribution of abundance in the community, then by more intensive sampling we should be able to find more of the increasingly rare species, as well as more individuals from the common species. Eventually it should be possible to sample every organism in the community and unveil increasingly rare species that are represented by only one or two individuals. Unfortunately, the huge amount of sampling that would often be required might ultimately prove more destructive of the community than informative to the scientist.

Only rarely can we put together complete abundance data for an entire taxon in a defined geographical region. The British Trust for Ornithology has assembled such a complete survey for the bird species living in England, Scotland, and Wales. The graph on this page illustrates the observed pattern of relative abundance plotted on a scale of logarithms to the base 2 (so that moving from left to right each bar represents a doubling of abundance). The population sizes of bird species in Britain range over six orders of magnitude; two species are represented by a solitary breeding pair (redbacked shrike and brambling), while several species have populations in excess of 5 million (chaffinch and wren). As all the birds in the country have been sampled, we can see a complete bell-shaped curve.

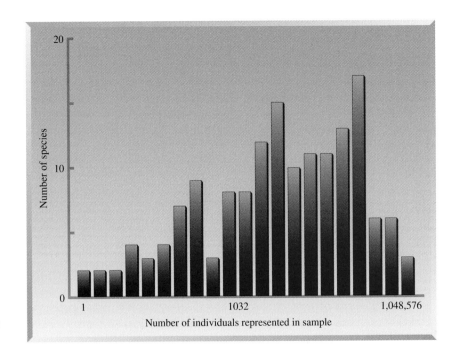

► The distribution of species abundance for all the bird species in mainland Britain. The figure is plotted on a scale of logarithms to the base 2; thus each division of the x-axis corresponds to a doubling of abundance. More than half the species have populations in excess of a thousand.

The pattern of relative abundance represented by the bell-shaped curve appears with remarkable consistency when we examine data for a variety of other taxonomic groups in a wide range of habitats. It is the same pattern that we would obtain if we were to take a stick and sequentially break it into sections at random points along its length. It is disconcerting and yet in some ways exciting that the relative abundance of species in any ecological community tends to conform to a relatively simple underlying pattern. The pattern provides us with a potential technique for estimating the total diversity and relative abundance of species in a community without having to sample so intensively that we effectively destroy the community: we need only sample enough to hint at the shape of the curve; filling in that shape will automatically give us the information on diversity and abundance that we seek. Even quite artificial distributions develop this pattern of abundance, including the tree species in Central Park in New York and those in Hyde Park in London. The major exceptions are the communities in agricultural fields, which are dominated by a single variety of a single species of grain, except for occasional outbreaks of its pests and pathogens.

It seems logical to ask whether any particular characteristics of species, such as physical attributes, determine which species are common and which rare. One popular hypothesis is that a relationship exists between the numbers of individuals and their body size, and that the larger species should be rarer. If there are again common underlying patterns shared by different groups of species, this may provide another potential technique for estimating the total abundance of species without having to physically count them all. To explore this hypothesis, we again examine data for bird species in the British Isles.

The graph on the next page, produced by Sean Nee and colleagues at Oxford University and the British Trust for Ornithology, illustrates the relationship between body size and abundance for 147 species of British breeding birds. Although the plot suggests that, as expected, abundance decreases with body size, the relationship contains a lot of scatter, with many small birds having small population sizes. It is possible that these rare species of small body size are specialized to live in rare habitats and simply do not have the space to sustain a large population. Indeed, it would be intriguing to see if habitat specialists, or species with narrow geographic ranges, consistently had smaller populations than species with broader ranges and more flexible habitat requirements.

What could explain this pattern of small-sized species having large populations and large-sized ones having small populations? Not surprisingly, large animals have to eat more, and the more food an animal needs, the larger the area of land through which it must forage to find enough nourishment. Thus the same plot of land will support many more small individuals then it will larger ones.

▶ This graph suggests that, for British inland breeding bird species, population size, or abundance, decreases with body weight, although there is a lot of scatter in the data points.

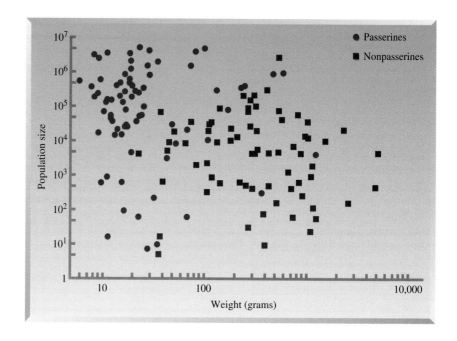

The relationship between abundance and body mass has also been examined for beetle species living in the canopies of 10 trees in the lowland rainforest of Borneo. In contrast to the British birds, many species are represented by a single individual. It is not clear how many of these loners are "tourists" that have wandered away from their usual habitat, where they may live at much higher population densities, rather than individuals from species that actually do live at low population densities. Furthermore, although abundance still tends to decrease with increasing body size, it does so at a slower rate. Nevertheless, both studies suggest that, within a defined area, larger species tend to exist at lower numbers.

The knowledge that we can gain by examining diversity at a single site is limited. For example, the question raised by the beetle study, whether species that are rare at one site are actually common somewhere else, can only be answered by broadening our survey to examine trends in diversity over much larger areas. Eventually, we may be able to answer a more general question: What determines the number of species in a region?

On our way toward an answer to that question, we first ask why so many species are rare. Are rare species long-time residents of the planet that are declining in numbers, or are they recently evolved species that are only just beginning to increase in numbers? Indeed, are some types of species always rare, and can we classify different types of rarity? A good way to begin our attempt to find answers is by examining the world's most completely known flora.

Geographical Patterns of Rarity and Abundance

Enthusiastic naturalists have been classifying the flora and fauna of the British Isles for over three hundred years. The collecting trips of the eighteenth century were organized with the same mixture of anticipation and pioneering spirit that characterize present-day collecting trips to the world's few remaining isolated areas. Although many of the swamps, woodlands, and heathlands that were the original collection sites have now been converted to farmland and shopping centers, Britain still has the best catalogued fauna and flora on earth. Even so, the data needed to quantify patterns of rarity are available for only 177 of the 1822 native plant species.

Deborah Rabinowitz, of Cornell University, and her colleagues have used these data to classify species of wild flowers using three parameters, each reflecting a different definition of "rareness." First, they classified the geographical distribution of a species as either wide or narrow: Could the species be found throughout most of Britain or was it confined to a small region? Then they examined habitat specificity: Was the species found in a broad variety of habitats or restricted to a single type of habitat? Finally, they noted whether the species was abundant in at least one location or whether it was everywhere very low in population. Each of the three classifications points to a different kind of "rareness." In fact, Rabinowitz classifies as "common" only those species that satisfy all three criteria—those that are found throughout Britain, exist in several types of habitat, and are high in number. Because the three dichotomies produce an eight-way

Rarity and Abundance in British Wild Flowers

| Population size | HABITAT SPECIFICITY | | | |
| | BROAD (66) | | RESTRICTED (94) | |
	Wide distribution	Narrow distribution	Wide distribution	Narrow distribution
Somewhere large (149)	58 *36%*	6 *4%*	71 *44%*	14 *9%*
Everywhere small (11)	2 *1%*	0 *0%*	6 *4%*	3 *2%*

In each case the upper figure is the observed number of species, the lower figure (in italics) gives the percentage of the total sample.

classification of species distributions, we are left with one type of commonness in addition to seven types of rarity.

Rabinowitz found that most species are abundant somewhere (149 versus 11); and similarly, that most species have a wide geographical range (137 versus 23). Although most species are not rare in these two senses, the majority of species are rare in the sense that they are restricted to a single type of habitat (94 versus 66). And indeed, if we consider the eight possible combinations of categories, the species of the commonest type (44 percent), are those that combine large population sizes, wide geographic ranges and restricted habitat specialization. Most of these are "rare habitat" specialists that live in marshes or sand dunes or on forest floors. The next most abundant category are the species that we would conventionally consider to be "common" species; yet only 36 percent of species in the sample fulfill this definition. Of the remaining six categories, the most frequent are the "endemic rarities": those species that specialize in one type of habitat, in one small geographic area, but are abundant there. Finally notice that one category is not represented at all: small populations in a variety of habitats, with a narrow geographic range. Is the absence of species exhibiting this or other types of rarity a real phenomenon? To answer this question we need to apply the same technique to other organisms.

A comparable analysis of rarity and abundance has been undertaken by Michael Reed, of the University of Nevada at Reno, using data for neotropical migrant bird species in North America. These are birds that breed in North America, but migrate to Central and South America for the fall and winter. Reed considered the distribution of the birds in both their breeding and winter ranges. In contrast to plants, bird species with narrow

Rarity and Abundance in North American Neotropical Migrant Birds

| Population size | HABITAT SPECIFICITY | | | |
| | BROAD | | RESTRICTED | |
	Wide	Narrow	Wide	Narrow
Somewhere large	59%	0%	9%	2%
	66%	*10%*	*7%*	*2%*
Everywhere small	16%	9%	0%	7%
	5%	*5%*	*0%*	*5%*

In each case the upper figure is for the breeding range; the lower figure (in italics) is for the winter range.

geographical ranges, broad habitat requirements, and low population densities were found. However, bird species with a restricted habitat range, wide geographical distribution, and low population density were absent. Furthermore, bird species appear considerably less specific than plants in the variety of habitats in which they breed.

Both these surveys suggest that some species are naturally rare, partly because they are restricted in their habitat requirements, partly because they may not have successfully colonized all available habitat patches. The most important message that these surveys present is that because species can be rare in different ways, we will need a variety of ways to accurately count and conserve rare species.

Geographical Gradients of Biodiversity

These patterns of rarity and habitat specialization are much too subtle and complex to visualize easily. A much stronger geographical pattern is clearly observed when we plot the numbers of, say, bird or tree species on a map. Unfortunately, for most species, we know only the general location where they may be found (or, worse, where the museum specimens were collected), and not much else about their ecology. Nevertheless, this information can tell us something about the geographical variations in species diversity. The map on the next page illustrates the diversity of bird species living in each square defined by a 10 degree shift in longitude and a 10 degree shift in latitude on the continent of North America. The diversity of bird species increases by almost a factor of 10 as we move from arctic areas to tropical areas. Mammals, too, show this distribution pattern.

Tree species diversity for North America follows a similar pattern: the number of tree species again increases by a factor of almost 10 as we move from the arctic toward the tropics. The map on page 21 also provides more detail on where we can find the peaks of tree species diversity. In particular, notice that the mountainous areas in the middle of the continent have much lower diversity than the areas toward the coast. Similarly, desert areas in the Southwest are relatively poor in tree species when compared with temperate and coastal regions. Where a generous supply of rain and sunlight encourages the lush growth of vegetation, as in the Southeast, the number of species is much higher. In general, tree species diversity increases as the climate becomes warmer and more humid.

The geographical variation that we observe in bird and mammalian diversity mirrors the underlying variation in tree species diversity. The similarities suggest that if we identify areas that are rich in one form of

▶ The contours represent the number of breeding bird species found at each location in continental North America. The diversity of birds is seen to increase by a factor of at least 10 between the arctic and the tropics.

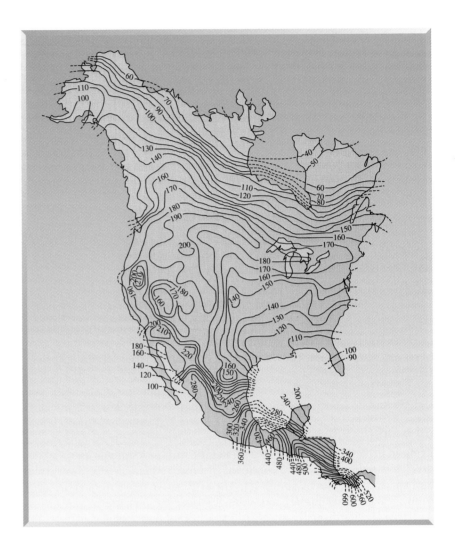

biodiversity, those areas are likely to be rich in other forms of biodiversity as well. All three groups reveal a considerable increase in diversity as we move into the tropics. Four major hypotheses have been proposed to explain the richness of species toward the equator. Unfortunately, the hypotheses are not mutually exclusive, and although one or another has been favored at any particular time, none has been shown to be completely wrong. It is likely that each hypothesis casts some light on the geographical trend for higher diversity in the tropics.

The first hypothesis simply acknowledges that, in contrast to much of the present temperate zone, the tropics have never been covered in glaciers during the ice ages that have periodically transformed the earth's climate.

During these times areas as far south as present-day Washington, D.C., London, and even Tokyo have been covered in ice. Whereas few forms of life could exist on a giant ice sheet, life in the tropics could continue to evolve and diversify. When the climate warmed and the glaciers retreated, then plants and animals could recolonize the areas once covered by ice, but this colonization is a slow process. Animal species would often have to wait to expand north until the more slowly migrating plant species had established themselves. Life in the north has simply had fewer long, continuous periods in which to evolve—this will tend to always produce fewer species as we move away from the equator.

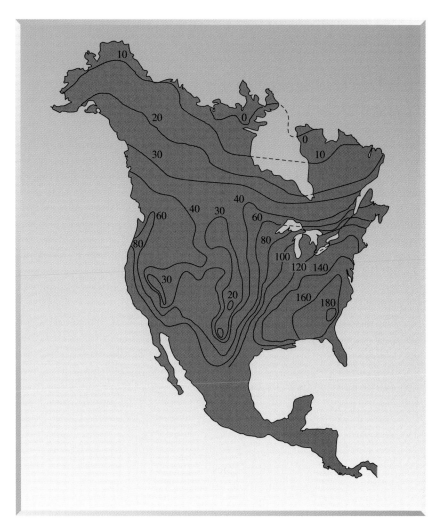

◀ Tree species diversity varies geographically and increases toward the equator. The tree species diversity was estimated by recording the number of tree species within the boxes of a grid covering North America. Each box covered 70,000 square kilometers. Contour lines were then drawn connecting boxes of equal species diversity.

The second hypothesis makes note of the fact that the majority of the world's land mass lies in tropical regions. The larger land area alone creates more opportunities for more species to exist in the tropical regions of the world. Third, the presence of more plant species in the tropics creates greater opportunities for the insect and bird species that feed on these species to specialize and diversify into more species. This process will tend to magnify levels of biodiversity in areas where plant diversity is high because of either of the two previously cited reasons. Finally, changes in climate and sea level, accompanying repeated cycles of cooling and warming, may have created isolated patches of forests in different parts of the tropics. If these periods of isolation lasted long enough, then isolated subgroups of once contiguous populations may have had the chance to evolve into new species.

Number of Species versus Physical Size

If, as we saw earlier, body size determines abundance for individual species, then perhaps it may also be an important variable determining species diversity. And indeed, we find an enticing relationship between mammal diversity and body size, whether we look at species in the British Isles alone or throughout the world. There are very few species of large mammals, increasing numbers of small to medium-size mammals, and, curiously, fewer species of very small mammals. The sparcity of species at the endpoints of these distributions perhaps reflects the extreme physiological constraints possible in mammalian species. For example, it is impossible for a terrestrial mammal to grow much larger than an elephant without dangerously overheating during periods of vigorous activity.

The ecologists Evelyn Hutchison and Robert MacArthur originally advanced arguments for why we should expect a simple scaling relationship between species diversity and body size. New species can evolve when isolated populations no longer interbreed. For this to happen, at some earlier point the original population must have spread out, then become separated from one or more of its outshoots. Large-sized species require more land to support a given number of individuals (and they can often travel longer distances than small-sized species), so the same area that could support two or three separate populations of a small species, and thus potentially give rise to two or three new species, may only support one population of large-sized animals, with no opportunity to diverge and form new species. Essentially, if we assume that the world is two-dimensional for terrestrial organisms, the possibility of new species of individual characteristic length, L, finding new regions to colonize may scale as $1/L^2$, (or $1/M^{2/3}$, where M is mass). The decline in diversity among larger mammal species indeed tends to follow this trend.

How general is this relationship between body size and species diversity? It is possible, by assembling information on all the species classified so far by taxonomists, to produce a diagram that illustrates the relationship between body size and number of species for all taxonomic groups of animals. The pattern that results is similar to that for mammals: as body size becomes larger, the number of species decreases at about the same rate (the slope of the dashed line in the diagram). Just as for mammals, there is again a pronounced reduction in the numbers of very small species within any group.

Why, no matter what taxonomic group we look at, are there few very small species? Tom Fenchel of the University of Copenhagen has examined this question for a number of marine and terrestrial groups. His synthesis suggests that smaller organisms within any group tend to be more cosmopolitan than the larger organisms and to occupy much larger ranges. Fenchel points out that microscopic algae, bacteria, and single-celled protozoans such as ciliates also live at very high population densities—one milliliter of seawater contains about a million bacteria and a thousand protozoa. Populations of these species almost never become fragmented, and genetic isolation is very rare. The appearance of a new species is thus a comparatively unusual event, and we commonly find the same species in completely different parts of the world.

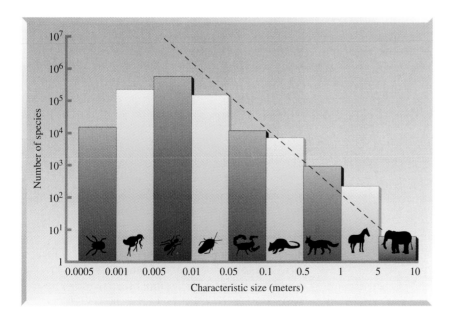

◀ A rough estimate of the number of all known terrestrial species, categorized according to characteristic body length, L. The dashed line indicates the relation $S \approx 1/L^2$.

▲ Even now, a new species of mammal is occasionally discovered. This species of tree kangaroo (genus *Dendrolagus*) came to the attention of taxonomists only recently, in 1994, when one of the animals was spotted by a dog accompanying the Australian mammalogist Tim Flannery and a local Dani hunter on a trek through the New Guinea rainforest. The indigenous people of the area had long known of the species, however.

The relation between body size and number of species breaks down for organisms less than one centimeter long. The tiniest organisms may produce new species only rarely, as just explained. But in addition, the data may be skewed by a sampling bias that exists because the taxonomic groups of organisms have been studied at a wide variety of intensities. For example, almost all bird and mammal species have now been discovered. The rates at which new bird and mammal species were found dropped rapidly once all the earth's major habitats and previously unexplored areas had been visited. In contrast, the rates at which we discover and describe invertebrates is considerably slower, partly because there are so many invertebrate species, and partly because there are fewer trained taxonomists in this area. The undescribed species that fill out the gaps in this distribution represent the greatest failing in our understanding of life on earth; we still do not know how many other species we share the planet with.

How Many Species Are There on Earth?

One of the first questions an extraterrestrial visiting earth might ask is, "How many species live here?" Ironically, the answer to the ultimate question in the quantification of biodiversity is astonishingly vague: we know only that there are somewhere between 1.5 and 30 million species of plants and animals sharing the planet with us. One recent estimate suggests that the true figure may be around 6 million, but the majority of species remain unnamed and unclassified by taxonomists.

Between 1.5 and 1.8 million living species have already been described by taxonomists: approximately 750,000 of these are insects, 41,000 are vertebrates, and 250,000 are plants. Invertebrates other than insects—fungi, algae, bacteria, viruses, and other microorganisms—make up the remainder of the described species. Although taxonomists have estimated the relative contribution of the different groups to the overall pattern of biodiversity, our picture is still very incomplete, due to a number of sampling biases. For example, a number of major habitats remain poorly explored, including the floor of the deep ocean and the tree canopies and soil of tropical forests.

Many fewer species of organism have been recorded from the oceans than from terrestrial ecosystems. The majority of marine forms that have

▶ Groups of organisms are ranked according to their relative contribution to total biodiversity. More than half of all known species are insects. In contrast, only 4000 of all known species are mammals and around 10,000 are birds, forming respectively 0.025 percent and 0.066 percent of the roughly 1.5 million recorded species.

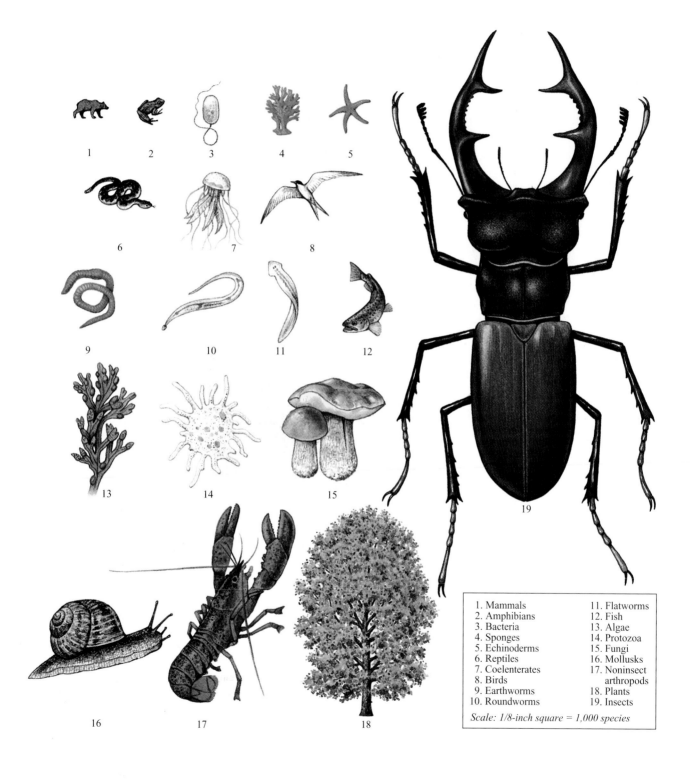

1   2   3   4   5

6   7   8

9   10   11   12

13   14   15

16   17   18

19

1. Mammals
2. Amphibians
3. Bacteria
4. Sponges
5. Echinoderms
6. Reptiles
7. Coelenterates
8. Birds
9. Earthworms
10. Roundworms

11. Flatworms
12. Fish
13. Algae
14. Protozoa
15. Fungi
16. Mollusks
17. Noninsect
    arthropods
18. Plants
19. Insects

Scale: 1/8-inch square = 1,000 species

▲ Entomologist Terry Erwin uses an insecticidal fog to sample the insects living in a tropical tree. The thousands of insects living in a single tree drop to the ground where they are collected on sheets surrounding the base of the tree.

been described are from inshore regions and coral reefs. The deep ocean has been considerably undersampled; some oceanographers estimated that until the mid-1980s the total area of deep ocean bottom sampled by biologists barely exceeded an area the size of a tennis court!

Yet levels of biodiversity in the deep ocean may be very high. When samples were taken along a 176-kilometer transect off the coast of the eastern United States, researchers identified a total of 798 species, from 171 families and 14 phyla, among a total of 90,677 individuals. As the researchers moved farther off shore, at first the rate at which they found new species increased rapidly. Eventually the rate settled down, and new species were located at a rate of around one species per kilometer of transect. By scaling this estimate up, and correcting for differences between the continental shelf and deep ocean habitat, a provisional estimate is obtained that suggests there may be as many as 10 million undescribed species in the world's oceans.

There appear to be differences in the hierarchical structure of biodiversity between marine and terrestrial environments. In the marine environment, there are many marine families and subphylums, with only a few species in each of them. The terrestrial environment, in contrast, is characterized by fewer subphyla, each of which contains many more species. If we compared the evolutionary trees of marine taxa with those of terrestrial taxa, the radiation of terrestrial species would resemble a tree with several large branches, each radiating into several smaller branches, with a great multitude of twigs at the ends. In contrast, the ancestral tree of marine species would contain many long thin branches, each segmenting into a few smaller branches.

The other largely unexplored areas of great biodiversity are the world's tropical forests. Although tropical forests cover only 7 percent of the planet's land surface, they contain a significant proportion of its biodiversity. Much of this diversity is dominated by insects, and particularly beetles. Entomologist Terry Erwin has attempted to quantify the numbers of beetles living in the canopy of rainforest trees, in the hope that his results would shed light on the total number of species in the rainforest and even the total number of species in the world. Erwin sprayed an insecticidal fog to knock down canopy-living species from one particular tree species, *Luehea seemannii*, in Panama. He repeated the procedure over three seasons and then extended his study to examine the beetles associated with *L. seemannii* in other parts of its range. He found scarcely any species of beetle shared in common among *L. seemannii* from three different parts of South America. More astonishing, he discovered a new species with every second or third beetle examined.

Erwin found over 1100 species from his original samples in Panama. Although he is still putting formal taxonomic identities to these species, he has initially classified them by lifestyle: the great majority (80 percent) were herbivorous, and of the remainder 5 percent were predators, 10 percent were fungivores, and 5 percent were scavengers. An estimated 20 percent of the herbivorous beetles were specific to *Luehea*, rather than living on many kinds of trees. This figure leads to a guesstimate of 160 species of canopy beetle specific to a typical tropical tree species. Erwin suggested that these data could be used to provide a rough estimate of the total beetle diversity in tropical forests. If we assume that beetles represent about 40 percent of canopy arthropods, then we would expect to find 400 canopy arthropods per tree species. If the canopy fauna is at least twice as rich as the fauna of the forest floor, then we would expect there to be 600 arthropod species specific to each tree species. And finally, if, as one widely accepted estimate states, there are approximately 50,000 species of tropical forest tree, then there may be as many as 30 million tropical arthropods!

Although it is easy to criticize each step of this calculation, it provides an important example of the type of study biologists will have to perform if we are to more accurately quantify biodiversity in the tropics. In particular, it should encourage similar studies on a wider range of tree

Erwin counted more than 1,100 species of beetle in *L. seemannii* canopies; he decided 160 were specialized to the canopies of that tree.

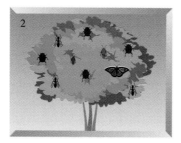

Beetles constitute 40 percent of all insect species, so Erwin calculated that 400 species of insect are specialized to the canopies of *L. seemannii*.

If two-thirds of all insect species live in the canopy, then there would be 600 insect species altogether specialized to *L. seemannii*.

There are an estimated 50,000 tree species in tropical forests, giving 30 million projected insect species (50,000 trees × 600 insects).

◄ Terry Erwin has attempted to estimate the number of insect species in the world by counting the number of beetles in the canopy of a single tropical tree species, and extrapolating from there.

species from different places and more detailed sampling of all phyla living in the tree canopy and in the ground under the tree. Finally, we need to examine the specificity of insects for different tree species, and we need more information on the actual diversity of tropical trees. Unfortunately, at current rates of tropical forest destruction, we probably lose 300 species of tropical tree each year. If Erwin's figures are correct, this represents a loss of 180,000 arthropod species each year, or around 500 species every day!

The rainforest's potential for containing enormous biodiversity has been demonstrated more recently by Hanna Tuomisto and colleagues from the University of Turku in Finland and the Instituto Nacional de Recursos Naturales in Peru. Tuomisto's group has used detailed ground surveys and satellite images to examine how biodiversity changes from spot to spot in the lowland Amazonian rainforests, at spatial scales ranging from a meter to hundreds of kilometers. They have found that the diversity of plants and animals varies enormously with topography and soil conditions. Their work strongly suggests that instead of thinking of the tropical forests within a region as one large homogenous mass, we should think of them as assemblages of many different patches blending into each other, each of which contains a different combination of plant and animal species. When Tuomisto's group examined forests from the ground, they found that each 30-kilometer transect contained an average of four different habitats. When they examined forests through satellite images, they found that each 34,000-square-kilometer section contained between 21 and 54 different types of habitat. The mean diameter of any patch of habitat was only 4.6 kilometers in the upland unflooded forest called terra firme and 1.7 kilometers in inundated forest. When we consider the number of beetle species found by Terry Erwin on a single tree species, it becomes clear that this mix of habitats may give rise to considerable biodiversity in the tropics. Certainly, if we are to conserve biodiversity, we have to learn more about the underlying geographical patterns in the way that plant and animal species are distributed.

Some Taxonomic Complications

Erwin's suggestion that there may be as many as 30 million species of insects produced a huge controversy. A collation of published and unpublished information from a survey of taxonomists in museums and universities leads to a substantially reduced estimate for the total number of species inhabiting the earth. This survey suggests that a figure on the order of 10 million insect species is more tenable, and one of around 5 million not improbable. The number, though lower than Erwin's estimate, is still spectac-

◄ Chimps and human beings share 99 percent of their DNA.

ular, and if nothing else emphasizes the amount of biodiversity we are losing before it is even classified.

Even these more careful estimates may be biased by differences in the relative effort expended on researching different taxonomic groups. In particular, there is a significant tendency to lump "species" together at lower taxonomic levels. For example, studies on the bacteria *Legionella pneumophila* show that organisms that share less than 50 percent of their DNA are still classified as the same species—a span as large as the characteristic genetic distance between mammals and fishes. In contrast, chimps and humans are classified as different species even though they share 99 percent of their DNA! At the other extreme, some taxonomists tend to endlessly subdivide the taxa they study. While some taxonomists see only around 20 species of the British blackberry (a real lumping taxonomist might only concede 3!), others see as many as 200.

Kevin Gaston and Robert May have pointed out that one further crisis thwarts our present attempts to quantify the number of species on the earth: ever fewer numbers of people are being trained as taxonomists. In the United Kingdom the number of taxonomic researchers has declined from 552 in 1980 to 514 in 1990, a 7-percent drop that almost matches the 6-percent decline of funding in this area. Similar figures have been obtained from recent surveys in the United States and Australia. More disconcertingly, taxonomists as a group are showing signs of aging. In 1980, 23 percent of teachers of systematic biology in the United Kingdom were less than 35 years old, while 43 percent were older than 46; in 1990, the corresponding proportions were 8 percent and 63 percent. It seems that not only biodiversity, but the principal class of personnel trained to quantify it, is in decline.

Furthermore, recent surveys indicate a major disparity between where trained biologists live and presumably work and where biodiversity is concentrated; about 80 percent of ecologists and entomologists are based in North America or Europe, and only 4 percent in either Latin America or sub-Saharan Africa. The distribution of taxonomists is almost completely the opposite to the best current estimates of where most of the earth's biodiversity is found.

Moreover, the majority of taxonomists are concentrated as specialists in groups for which the majority of species have already been described. In both Australia and the United States, around 32 percent of animal taxonomists work on tetrapods (reptiles, birds, and mammals), 6 to 11 percent on fish, 30 to 32 percent on insects, and 25 to 32 percent on other invertebrates. Gaston and May use figures like these to suggest that whereas in Australia there are around 3 undescribed species for each tetrapod taxonomist, there will be more than 400 undescribed species for each invertebate taxonomist.

Changes in Diversity through Time

Although biologists have not yet obtained a total estimate of the number of species that share the planet with us, they have set up some experiments to monitor the way in which the diversity in any one area changes with time. These experiments study plant species that live in agricultural landscapes, yet they provide important insights into the likely effect of human activities on the as yet unquantified diversity of species that live in the wild.

One of ecology's longest-running experiments is the Parkgrass experiment at Rothamsted in England. This experiment was set up in 1856 to determine the effect of different fertilizers on the grass yield of a meadow. A large number of plots of grassland have been monitored since the experiment started, some subjected to fertilization and others left unfertilized. The relative abundance of species changes dramatically in the fertilized plots: the plots start off with a wide diversity of species and only a few common ones, but gradually rare species go extinct and the common species become even more abundant. A hundred years after the experiment started, the fertilized plots are dominated by a few common species that are very aggressive in competing for nutrients. These experiments could be perceived as an allegory for the fate of many natural habitats in the twentieth century. Rare species are becoming extinct, and communities are being dominated by a few common species.

The opposite pattern is seen when agricultural land is abandoned: the land is soon colonized by plants from the surrounding habitats or by species whose seeds have remained dormant in the soil. In these abandoned fields diversity increases through time, and the community eventually develops a pattern of relative abundance that is similar to that observed at the beginning of the Rothamsted experiment.

I'm not convinced that an extraterrestrial visitor would be much impressed with human intelligence if we confessed that our estimates of global biodiversity are only accurate to within 5 million, and that based on some fairly tenuous assumptions we think between 5 and 10 million species share the planet with us, of which only 1.5 million have been described. Our estimates of global biodiversity are constrained by our ignorance of the number of species that live on the ocean floor and in the canopy of the tropical rainforests. Preliminary studies in these habitats suggest that the total number of species inhabiting the earth may be in excess of 20 million. As huge as these numbers are, most ecologists and evolutionary biologists believe that they could drop precipitously, and that we are now in fact in the early stages of a major decline. The Parkgrass experiment hints that human interference could be a possible cause. Using information on global patterns of diversity and its geographical distribution, the next chapter attempts to understand the factors that are contributing to the present-day loss of biodiversity.

Habitat Fragmentation and Loss

We had driven all day across the Gran Sabana, the vast plateau in southern Venezuela, where giant waterfalls cascade down near-vertical cliffs and where most of the plants eat insects to survive in the nitrogen-poor soil. Arriving after nightfall at a tiny village, we pitched our tents behind the ancient cathedral and thought about waking in the morning to watch the sunrise over the vast forests that lie between the Orinoco and Amazon rivers. As the morning mist cleared, the forest emerged, not as a continuous sea of vegetation, but as a broken mosaic of forest and patches of clearings. From the more distant clearings columns of smoke arose that mixed with the mist; the nearest clearing, at the bottom of the cliff beneath us, was filled with neat, regular rows of plants. Following our guide down the cliff path and into the "jungle," we quickly discovered that the edge of the forest was slowly being replaced by fields of manioc and pineapple. Gradually, but inexorably, the world's largest continuous tropical forest is being converted to small patches of agriculture.

◀ Formed of some of the oldest rock on earth, the Gran Sabana plateau harbors some of the planet's most exotic species, including the largest insect-eating plants. Yet human settlements are steadily eating away at its forest habitats.

33

▶ A satellite photo of the Amazonian rain-forest in Brazil shows the impact of a major highway. Roads branching off from the highway into the forest in turn give rise to secondary and tertiary paths that fragment the forest into small, irregular patches.

The scenes we witnessed in Venezuela are a variation on a theme that is repeated every day in many parts of the world. The single process most responsible for the present high rates of loss of biodiversity is habitat destruction. Throughout the world key natural habitats, whether forests or savannas, heathlands or swamps, are being converted to agricultural land, urban areas, or lifeless deserts. Although humans have been transforming the landscape throughout our history, the human population surge in the twentieth century has rapidly increased the rate at which natural areas are converted to biologically depleted areas in which we live, work, or farm. The communities of species that live in these natural areas are thus disappearing at rates that roughly match the rates of expansion of the human population.

As forests are converted into agricultural land two processes lead to a reduction in biodiversity. Habitat loss per se obviously decreases the area of land available for wildlife. However, habitat loss rarely consists of the simple paring away of the edge of an area of pristine habitat. Usually it is compounded by the fragmentation of the original contiguous area into an increasing number of smaller fragments, each of which may be insufficiently large to support viable populations of all the original inhabitants.

The Destruction of the Amazon Rainforest

The Amazon basin in Brazil provides a classic example of the way that rainforest erosion proceeds. Initially, the construction of the Belem-Brasilia highway in 1958 provided a thousand-mile pathway through a previously uninhabited area of rainforest. The highway allowed the development of small farms and villages along its path, and these in turn have given rise to a system of secondary roads. As these roads branch off into the forest, they provide access to new areas that are then developed for agriculture—mainly pasture for cattle. To prepare the land for pasture, all but the largest trees are cut down (giant strangler figs are often left to provide some shade) and then burned to provide a rapid release of stored nutrients, which feeds an initial "green flush" of vegetation.

▼ Cattle graze where rainforest once stood in the state of Rondonia, on Brazil's western frontier. Rondonia has been inundated by a huge "land rush" of settlers since it was discovered to have better-than-average soils in 1974. Its population has increased by over a thousand percent, and its forests have been devastated.

In the twenty years since the opening of the eastern Amazon, cattle have increased in number from essentially zero to around 5 million. Unfortunately, in the nutrient-poor soils of most tropical forests, these "slash-and-burn" agricultural practices produce cattle pastures that last for only two or three seasons. In many cases the nutrients released from the burnt vegetation are quickly used up by the rapidly growing agricultural plant or animal species, and this loss is compounded by the washing away of nutrients in tropical downpours. The farmer and his family then have to move on and create a new pasture for their cattle by cutting and burning a new area of forest. In Brazil, this practice was subsidized in the 1970s and 1980s by the government, which provided tax incentives for cattle ranching in an attempt to encourage migration away from the overpopulated, urban, and drought-ridden regions in other parts of the country. This policy drove the destruction of large areas of Brazilian forest (as many as 450,000 square kilometers) until international pressure convinced Brazil to remove the tax subsidies in the late 1980s.

Similar processes are taking place in the tropical forests of other South American countries and in those of Central America, Africa, India, and the Far East. In many of these areas timber is first removed to supply the demands of the world timber and wood pulp trade. The roads constructed to remove the timber then allow the subsequent arrival of a transient farming community, who clear the remaining land by burning what is left of the forest. The natural habitats and wilderness areas being converted to agricultural land are not limited to rainforests; savannas, heathlands, and coastal wetlands are all being converted to land for human occupation and exploitation. Of course, the need for humans to take land for cities and agriculture is nothing new. The major difference between the current rates of habitat destruction in the tropics and the historical rates of habitat conversion in Europe and the United States is that the rate has become much faster in the late twentieth century.

We observe similar patterns of habitat conversion if we compare forested areas in central Europe between A.D. 900 and A.D. 1900 and in the United States between A.D. 1620 and A.D. 1920. In both continents extensive areas of forest have been transformed to agricultural land in order to provide food for expanding human populations. In the United States there is an almost exact match between the rate of forest loss and the increase in cropland. In Europe natural habitats were converted to cropland at a rate of around 0.1 to 0.3 percent per year, over the course of 15 centuries; in the United States, the rates of conversion have been somewhat higher— rough estimates suggest figures in the range of 0.7 to 1 percent per year.

The modern rates of land conversion in the tropics equal the highest rates ever seen in the United States, and in some places far exceed them. At the present time 1 percent, or 10.5 million hectares, of tropical rainforest are lost each year, an area roughly equal in size to the state of Connecticut or the country of Costa Rica. Such estimates can be used to calculate the time until a fixed proportion of the habitat is converted to either agricultural land or desert. The estimates suggest that the majority of tropical rainforests will completely disappear by the middle of the twenty-first century. Yet even these estimates may be optimistic, because the rainforest is being destroyed at rates faster than the average in countries such as Brazil

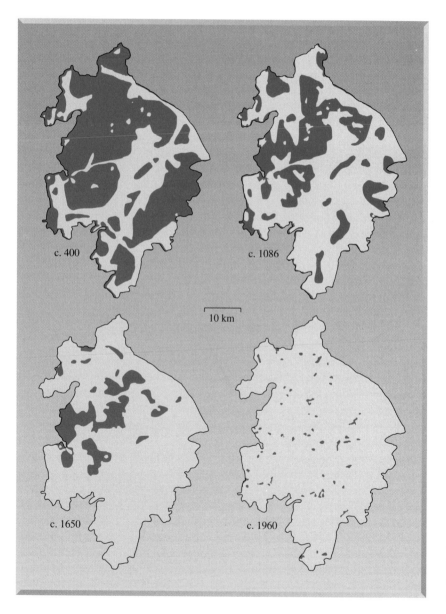

c. 400

c. 1086

10 km

c. 1650

c. 1960

◀ Between A.D. 400 and A.D. 1960 the forests of Warwickshire, England, became increasingly fragmented and eroded, until only a scattering of tiny patches remained.

► Where once huge forests stretched over half the continental United States almost without interruption, there remained by 1920 a rashlike spread of isolated patches. Each dot in the map for the year 1920 represents 25,000 acres.

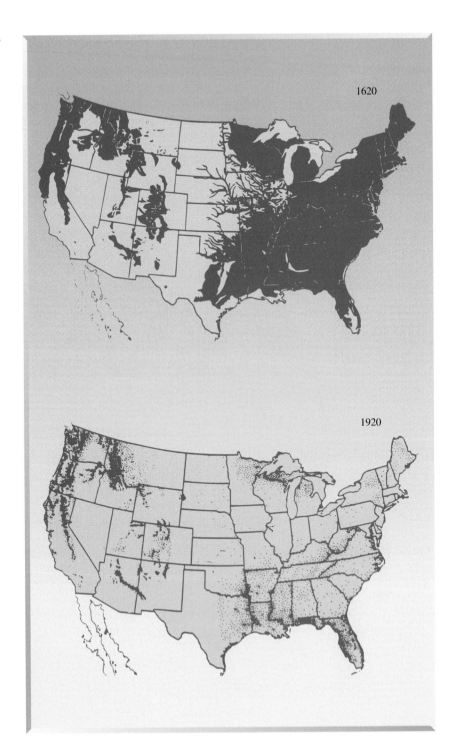

1620

1920

(2.2 percent), India (3.3 percent), and Myanmar (2.1 percent). In particular, the forest of Madagascar has been reduced to perhaps 2 percent of its original cover, and the coastal forest of Brazil to perhaps less than 1 percent of its original cover. Altogether, approximately 40 percent of the land on earth that once supported tropical rainforests has now been converted to some other type of habitat—primarily because of human actions.

Not all types of forests are destroyed at equal rates. The soils of forests classified as dry or moist are much better for raising crops than the thin soils of rainforests. Most of these dryer forests have already been converted to cropland by human populations living in the area. In contrast, until the last few decades wet forests were subjected to lower rates of deforestation, particularly when they stood in inaccessible locations. Nevertheless, now that the supply of readily accessible forest has declined, attention has switched to wet tropical forests as the final source of agricultural land, and of lumber as well. The arctic forests of Canada and Russia are another example of a previously ignored forest that has become attractive as a source of lumber. Ironically, developments in the paper-making industry that have increased the growth of recycled paper have also made it possible to use spruce and conifers as a prime resource for paper and wood pulp. Once considered of little value, the arctic forests of Canada and Russia face a whole new threat.

The remaining wet tropical forests of South America, Africa, and the Far East are the world's largest source of biodiversity. The late botanist Alwin Gentry, of the Missouri Botanical Garden, found that as the areas he

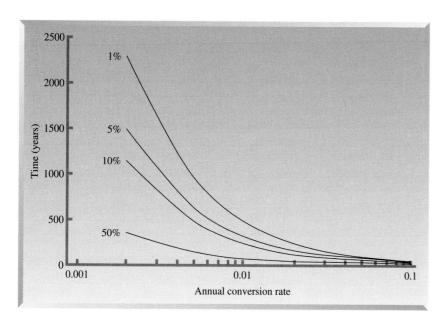

◀ For different constant annual rates of habitat conversion, we can calculate the time some proportion of the original habitat is converted into cropland or uninhabited land. The contour lines show the time until 50 percent, 10 percent, 5 percent, and 1 percent of the original habitat remain. At a constant rate of habitat conversion of 0.01, it would take fewer than a hundred years for 50 percent of the original habitat to disappear.

examined of South American forest increased in rainfall from 1000 to 3000 millimeters per year, the number of tree species found on a 0.1-hectare plot increased from around 40 to 140 species. The plant community gained about 50 tree species for every 1000 millimeters of rainfall; thus, tree species richness approximately doubles from dry to moist forests and triples from dry to wet forests. As the plant diversity increases, the more varied the resources that become available for animal species, and we see a huge increase in the animal diversity of a region as well. When we destroy these regions to create new agricultural land, we lose more biodiversity than is lost when we convert any other type of habitat.

Human Population Growth and Habitat Loss

If humans have one ecological feature that distinguishes them from all other species, it is their ability to alter the landscape in which they live. The only other species that make such significant changes to ecosystem structure and function are beavers, when constructing dams, and leaf-cutter ants, when defoliating certain trees and shrubs around their massive nests. Like beavers and leaf-cutter ants, humans modify their habitat in order to increase its value to them as a resource. The conversion of wilderness to agricultural land, or other forms of human use, is fundamentally linked to human population expansion and economic development. Obviously there are many subtleties in this process; nevertheless, it is hard to escape the basic fact that an increasing human population requires larger areas of cropland to provide food, as well as areas in which to live and process the resources that make human civilization viable.

Although the growing human population over the last two centuries has managed to crowd itself into an increasing number of huge cities, the land needed to feed these people has increased at a rate that is a simple function of human population size. Increases in agricultural efficiency are the only real factors that can distort the simple linear relationship between the size of the human population and the area of land required to feed that population. Agriculture has become considerably more efficient through technological innovation, the application of fertilizer, and the development of high-yield strains of cereal crops. Yet increasing evidence suggests that we have already maximized yield per unit area of land to the extent that we can. In the face of increasing fertilizer costs, reduced pesticide efficiency, and a growing human population, the demand to convert wild lands for agriculture continues to rise.

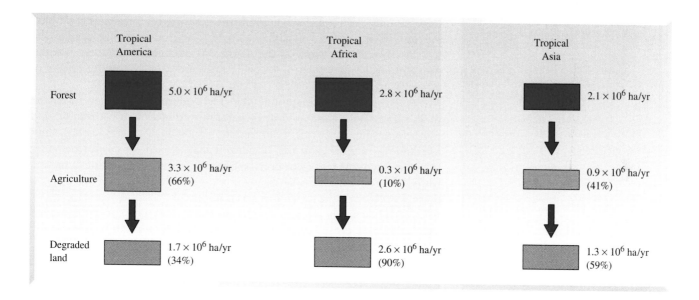

Looking across a large forest that is slowly being eroded by small patches of agriculture is a sobering experience, but only by examining maps of forest cover or sequences of photographs taken from space every year for decades can we fully appreciate the scale of forest fragmentation and destruction. These regular updates via satellite add a further twist to the tale. Although the major agents of forest destruction are farmers clearing land for either shifting or permanent cultivation, the increase in land under agriculture does not match the decrease in forest area. For the entire tropics the expansion of croplands and pastures accounts for only 27 percent and 18 percent respectively of total deforestation. The remaining 55 percent of land lost from forests is only matched by the increase in "other" lands. Although a small percentage (1 percent) of the increase is accounted for by expanding cities, the rest is degraded land—abandoned pastures and croplands. Unfortunately, land conversion often has unintended side effects that ultimately reduce the economic yields of crops and make the land uninhabitable. Erosion, reduced rainfall, salination, reduced soil capacity for water, and increased frequency and severity of floods have all come to plague land that was once optimistically put to the plow. Once degraded, the land may take between 50 years and 10 centuries to recover some resemblance of its original vegetation cover.

Some of these unintended effects are limited in their reach, but others extend well beyond the boundaries of the area converted. For example, recent estimates suggest that around 25 percent of the greenhouse gases being released into the atmosphere are from the conversion of tropical forests

▲ Between the years 1980 and 1985, tropical forests were transformed into agricultural lands, and agricultural lands into degraded wastelands, at the average annual rates indicated. The size of the boxes, representing the net loss of forests and the net gain of agricultural and degraded lands, illustrates that in tropical Africa and Asia far more wasteland has been created than cropland and pasture.

to pasture by slash-and-burn agriculture. These greenhouse gases are having long-term effects on global climate that may radically change the weather in different parts of the world. We will return to these broader problems in chapter 9; here we will concentrate on the direct effects of habitat conversion.

At present approximately 4700×10^6 hectares, or 23 percent of the earth's land surface, are used in agricultural production. This area of land can be coarsely divided between the 70 percent that is permanent pasture for domestic animals and the 30 percent that is used for crops (including trees). Ecosystems that can never be used for intensive agriculture, such as rocks, ice, tundra, and deserts, comprise around 4440×10^6 hectares; if these are added to the small area taken up by the world's cities (less than 1 percent), unfarmable ecosystems amount to about 31 percent of the world's land area. This means that somewhere between 46 and 60 percent of the earth's surface that could be used for crops is currently under cultivation. Although this figure suggests that we might be able to double our current productivity by converting all appropriate remaining wilderness areas to agriculture, this would only be true if the land remaining is as suitable for farming as the land being farmed at present. This assumption is at best naive and in many ways deeply flawed. Humans have tended to colonize the richest agricultural regions first and have left the poorer regions uninhabited. Thus much of the currently unused land would probably produce crops for only two or three years. Its conversion could also damage agricultural productivity in other regions. For example, computer models of the Amazonian climate indicate that if large areas of tropical forests were replaced with degraded grass pastures, rainwater that is now captured by trees and evaporated to form the cloud cover would instead be washed away in rivers to the sea (taking much of the soil with it).

At present 38 percent of the agricultural land in tropical Africa, 38 percent of the agricultural land in Latin America, and 67 percent of the agricultural land in tropical Asia have been derived from the conversion of

▼ Costa Rica has been rapidly deforested in the years since 1940.

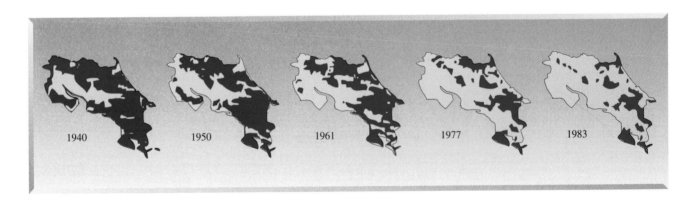

1940   1950   1961   1977   1983

forests in the last 140 years. As a consequence, 28 percent of the Latin American forest has been lost and between 34 and 38 percent of the South and Southeast Asian forest. To a good approximation, the land converted in the last 140 years has more than doubled the amount of land converted since the agricultural revolution began 10,000 years ago. In the tropics that doubling has taken place predominantly in the last 40 years. These two trends closely match the changes in the human population, which, after requiring around one thousand years to double in size for most of recorded history, doubled from 1 billion to 2 billion in the 125 years before 1925, and doubled again to 4 billion during the following 50 years.

This rapidly expanding human population has led, directly and indirectly, to rates of habitat loss that are now particularly alarming for the world's most critical storehouse of biodiversity—tropical forests. As this natural habitat and others are fragmented and reduced in size, normal ecological interactions are transformed in ways that lead to reductions in biodiversity. Through computer models on the one hand and field studies on the other, scientists are learning to predict the course of species losses—which species are most at risk and how fast species will disappear.

Vanishing Coastal Lands, Expanding Deserts

Tropical forests catch our attention because of their extraordinary, though largely unexplored, biological richness; nevertheless, they are not the only habitat that is threatened by the expanding human population. For example, coastal zones—which make up about 8 percent of the earth's surface (about an Africa and a half)—are one of the most threatened habitats on earth. They include coastal forests and marshes as well as watersheds that in some cases stretch quite far inland. Home to more than 50 percent of the human population, coastal areas are probably the most productive places on earth, for they supply 90 percent of the food resources to the huge body of humanity living there.

In many countries wetlands have disappeared when river water has been diverted for agricultural use, or to supply large cities. Perhaps the most dramatic example is the Aral sea in central Asia, until 1960 the fourth largest freshwater lake in the world. The diversion of waters from the two major rivers that supply the Aral sea has cut its input of water from 55 billion cubic meters of water a year to an average of 7 billion cubic meters. Some 20 of the lake's 24 fish species have disappeared, and the fish catch, which in the 1950s totaled 44,000 tons and supported 60,000 jobs, has dropped to zero. A similar fate has already met the now dried-up estuary of the Colorado river in the Gulf of Mexico. Blocked by dams and diverted to cities and farmlands, the flow of the Colorado into the sea has dropped from an average 20 billion cubic meters per year in the early years of this century to almost zero in all but flood years.

The encroachment of deserts into agricultural lands is an unplanned-for but often inevitable consequence of overexploitation of agricultural land. The most vivid example is taking place in central Africa, where the Sahara desert is encroaching southward into the north of Senegal, Mali, and Niger. The southward expansion of the Sahara intensified in the 1960s when pastoralists were provided with year-round water supplies for their cattle so that they might abandon their traditional migratory way of life and settle down. An important result was that cattle grazing intensified around the water holes. All plant material was eaten during the wet season, and no food was left for the longer dry season.

The perennial grasses on which cattle fed have deep roots reaching down to the water level. These grasses respond to increased grazing by re-sorbing their roots to produce more leaves. After a prolonged period of overgrazing, their roots can no longer reach the belowground water supply, and the plants are displaced by short-lived unpalatable plant species. These shallow-rooted species are less capable of holding the soil together, and the trampling of the cattle around the water holes leads to soil erosion and the creation of small patches of desert. To obtain new grazing land, the herders must cut down forest farther toward the south. The conversion of forests

▶ The depth of the water level in the oasis of Timbuktu in Mali, visible here as the depth of the well, poses a challenge to plant roots that won't be met if the plant must divert its growth to replacing leaves consumed by cattle.

to pasture is inexorably followed by desertification, further speeding the southern expansion of the Sahara desert.

Although some ancient deserts can support a diverse array of species, the low nutrient levels in deserts ensure that populations of these species exist at very low densities. Most of the plant and animal species that live in deserts have evolved special adaptations to exploit resources such as water and nutrients that may be available only a few times a year. The chuckwalla lizard of the southwestern United States desert, for example, feasts on spring vegetation—its source of both moisture and nutrients—that has been nourished by winter rains. When the vegetation withers, by early June, the lizard stops eating altogether. Like many desert species, it has evolved a low metabolic rate to conserve resources in the harsh environment. When a chuckwalla stops eating, it stops almost all other activity as well. Its body temperature falls and its rate of metabolism slows: it breathes more slowly and loses less water from its respiratory passages.

Desert species like the chuckwalla, with their slow metabolisms, reproduce slowly; their populations increase and expand at a very slow rate. These species do not respond quickly to change. Although the total area of desert in the world is increasing, the new areas are colonized by desert species only very gradually. In the deserts of North Africa, for example, areas disturbed by military activity fifty years ago have not yet returned to their previous state: tank tracks from World War II are still visible.

Insights from Islands

As forests are converted to farmlands or wetlands to industrial estates, the previously contiguous areas of natural habitat become fragmented into smaller patches whose number increases as their total area diminishes. In some ways the patches of original habitat can be considered islands in a sea of less hospitable terrain. Might they have features in common with islands of the oceanic kind? Ecologists and evolutionary biologists have long been fascinated with oceanic islands; they operate as isolated natural laboratories where ecology and evolution can be studied in communities that are simpler and less diverse than those present on larger islands and continents. How much does this knowledge of island biology apply to habitat "islands" that have been created on a much faster time scale?

In 1965 Robert MacArthur and E.O. Wilson proposed their theory of island biogeography to explain the patterns of species diversity they had observed on oceanic islands. The theory was based on a simple assumption—namely, that the number of species present on an island would ultimately be determined by the number of species colonizing the island, usually arriving from the nearest continents, and the number of species

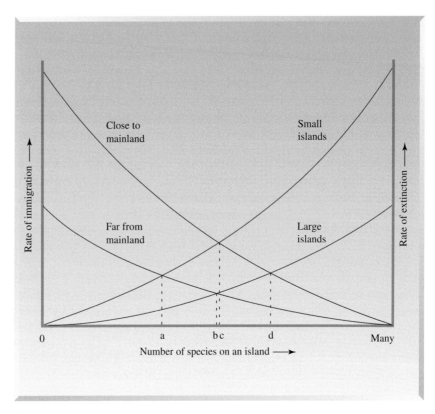

▲ The island biogeography model describes the relationship between the rates of immigration and extinction on oceanic islands. Islands near to a mainland source of potential colonizers will experience higher immigration rates than those farther away. As the number of species on the island increases, the rate of immigration declines, since there are fewer species left to colonize the island. In contrast, rates of extinction increase as the number of species on an island becomes larger. Small islands will experience larger rates of extinction than larger islands because their smaller populations are more vulnerable to extinction. Size and distance interact to produce different island communities, represented by points *a* (the number of species on a small island close to the mainland), *b* (the number of species on a small island far from the mainland), *c* (the number of species on a large island far from the mainland), and *d* (the number of species on a large island close to the mainland).

disappearing from the island, and becoming locally extinct. That is, the number of species depends on the balance between the rate of colonization and the rate of extinction. MacArthur and Wilson predicted that extinction rates would be lower on larger islands, which would have the area necessary to support larger populations of any individual species. Large islands would thus be expected to maintain larger numbers of species than smaller islands; by similar logic, islands that were a long way from a continent would contain fewer species than islands close to a mainland source of potential colonizers.

The theory is logical, but we rely on studies in the field to tell us if it is true. In most studies, it seems to be the case that the diversity of fauna and flora increases from smaller to larger islands. This pattern has been observed in a wide range of studies on different taxonomic groups, from lizards and amphibians on Caribbean islands, mammals in African game parks, and birds and butterflies in woodlots throughout the world. The results are almost identical when studies examine the number of species inhabiting patches of habitat such as woodlots, rather than islands. The remarkably steady increase in species number with area is shown most clearly in the form of species-area curves such as those on this page. However, that larger areas will support larger populations is not the only explanation for their greater richness of life forms. In addition, larger areas will contain a larger variety of habitats, and each type of habitat will contain a number of species that are specialists.

The theory has two very important consequences for biological diversity. First, if the total area of any type of habitat declines, it will support a less diverse variety of species. Second, some species will be lost sooner than others, in particular, species that require larger areas of habitat to sustain a viable population. Most predatory mammals and birds require substantial areas of habitat in which to hunt and locate their prey. A single pair of ivory-billed woodpeckers, for example, may require between 6

▼ Left: A species-area curve shows how the number of amphibian and reptile species increases with the size of the island, in this case islands of the Caribbean. Right: A species-area curve for the number of butterfly species found in 22 English woodlands.

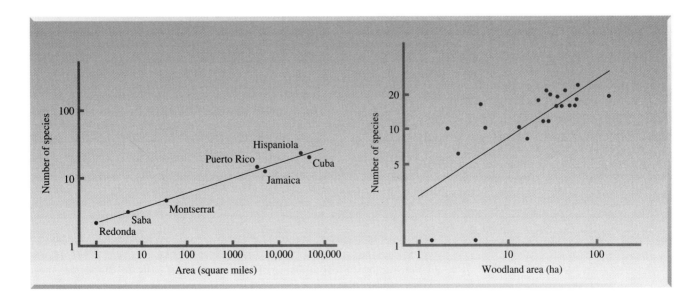

and 8 square kilometers of undisturbed forest, while the predatory European goshawk has a home range of approximately 30 to 50 square kilometers, and mountain lions may have home ranges in excess of 400 square kilometers.

Once a patch of viable habitat falls beneath these levels, it is no longer able to support even a single pair of individuals, and the species will go extinct in the patch. That extinction could in turn lead to a chain reaction of further extinctions. If predators such as mountain lions are no longer around to keep the numbers of herbivore species low, then herbivores such as mule deer may increase in numbers, overgrazing the land and leading to the extinction of some plant species and less competitive grazers.

Metapopulations

Because the species using a forest or heathland will include everything from soil microbes and nematodes through birds and large ungulates, it is likely that fragmentation will mean different things to different organisms. A patch of habitat that is a foraging patch for one species will include several home ranges of another. Similarly, a patch that includes several home ranges of one species will include whole populations of another species. In some cases, every member of some plant species will represent a patch of habitat for some insect species. A ripe fruit that represents a quick meal to a monkey or parrot can represent the entire habitat of several generations of fruit flies.

Interestingly, some species are well used to coping with small, and even disappearing, patches of habitat. Some types of habitat, such as swamps, forest clearings, serpentine grasslands, and ponds, naturally occur as fragments embedded in a mosaic of other habitats. Indeed, patches of woodland clearings and swamp may only be present for a short period before developing the characteristics of the habitat that surrounds them. Many of the species that live in these ephemeral patches of habitat have evolved life histories that include episodes of colonizing new patches of habitat. In some cases only the prereproductive young will disperse to new patches; in others individuals will move between patches at different times of the year, or between breeding seasons. A patch may become vacant of individuals from a particular species if its residents die, but new arrivals may subsequently take up residence there.

Populations distributed in this way are said to have a metapopulation structure. Many insect, and some bird and mammal, species have metapopulation structures. For example, the Heath Fritillary butterfly (*Mellicta athalia*) requires an abundance of food plants that grow in the warm, sheltered sites provided by well-drained soils in the first few years following the clearance of a patch of woodland. These conditions are provided by the English practice of coppicing woodland: the English have traditionally chopped back mature trees so that they would regrow with multiple trunks. After two to five years of growth the coppiced site becomes unsuitable, and the butterflies must colonize a new patch of habitat. Similarly Kirtland's warbler *(Dendroica kirtlandii)* nests only in young stands of jack pine, from 8 to 18 years old, growing in the barrens of Michigan after the occasional fire.

A population can persist as a metapopulation only when its members are not entirely confined to single patches, but make use of a conglomeration of fragments. Still, for the species to persist, patches of minimum crucial size must be present, and a minimum number of patches must be occupied at any one time. A patch that is too small or too far away may never be located by colonizers. Moreover, individuals of the species must be able to recognize which patches are suitable habitat. In some cases, the most vital clue is the presence of individuals of the same species in the patch. A patch that is uninhabited by others of the same species may never be colonized. When managing populations of these species, we cannot predict absolutely whether a patch of habitat is occupied by a particular species, but we can assign a probability that a patch of any particular size is occupied. That probability, known as the incidence function, will be determined by the colonization rate and by the birth and mortality rates of a population, just as the overall species diversity of an island will be determined by the balance between the immigration and extinction rates of individual species.

Larger animals, especially, are often endangered when the patches of habitat left, after human development has transformed an area, are smaller than their minimum home range. But fragmentation can lead to extinction through a number of other processes as well, some having a direct effect on the viability of a species, and others having more subtle, indirect effects. For an example of the latter, we turn to the vanishing American songbirds.

Where Have All the Birds Gone?

Ornithologists throughout the United States have become concerned about the overall decline in abundance of a large number of once common woodland bird species, especially east of the Mississippi. In some areas where bird populations have been monitored over the last forty years, the density of species such as the veery, black-and-white warbler, red-eyed vireo, and American redstart have decreased by as much as 50 percent. Other species, such as the Kentucky warbler and the yellow-billed cuckoo,

▲ This nest, belonging to the yellow warbler watching over it, has been parasitized by a brown-headed cowbird, which left the egg that hatched into the large nestling reaching toward the adult bird. The newborn cowbird is larger and more aggressive than its two warbler nestmates, and will manage to grab most of the food brought by the warbler parent.

have become locally extinct in woods where they were once known to breed regularly.

A number of mechanisms have been suggested to explain these declines. First, although the amount of habitat available to the species has remained roughly constant in the United States, most of these birds migrate to the tropical forests of Central and South America for the winter. There, much of their overwintering habitat has disappeared as forests have been converted to agricultural pastures. But the loss of overwintering habitats doesn't seem to fully explain the large drop in numbers experienced by many species. Some migrant species whose wintering grounds are intact still show a dramatic decline in population, while others remain unscathed despite the dwindling of their winter habitats. For a more complete explanation, biologists have turned back to the United States, where they have noticed that missing species can still be found in large continuous tracts of forest. Bird populations are most impoverished in smaller, isolated tracts. Forest fragmentation seems to be a factor in the decline of these birds, but, as it turns out, an indirect one. The root cause of the disappearances is actually another bird, the brown-headed cowbird.

Changes in the Abundance of Breeding Birds in Rock Creek Park, an Upland Deciduous Forest of Washington, D.C.[a]

Species (number of census)	1940s (N = 3)	1950s (N = 6)	1960s (N = 8)	1970s (N = 10)	1980s (N = 4)	Change in no. of pairs (1940s–1980s) (E = extinct)	Status[b]
Red-eyed vireo	41.5	39.2	28.1	10.6	5.8	− 35.7	M
Wood thrush	16.3	13.1	5.2	4.5	3.9	− 12.4	M
Black-and-white warbler	3.0	0.2	0.0	0.0	0.0	− 3.0(E)	M
American crow	2.0	0.7	2.0	0.9	0.6	− 1.4	R
Kentucky warbler	1.0	2.5	0.4	0.0	0.0	− 1.0(E)	M
Yellow-billed cuckoo	0.8	0.0	0.0	0.0	0.0	− 0.8(E)	M
American redstart	0.0	0.2	0.9	1.0	0.0	0.0	M
Brown-headed cowbird	0.5	2.1	4.4	0.5	0.8	+ 0.3	S
Veery	0.0	3.8	6.4	2.4	1.3	+ 1.3	M
Total species	26.5	24.0	24.0	20.5	24.5	− 2.0	
Density (pairs/100 ha)	90.1	82.4	87.1	43.2	34.9	− 55.2	

[a] Numbers represent mean number of pairs recorded per census during the 10-year interval. Not all birds censused are included in the table.

[b] Status: M = long-distance migrant; S = short-distance migrant; R = resident.

There has been a huge increase in the range and population size of the brown-headed cowbird *(Molothus ater)* over the last fifty years, largely as a result of changes in agricultural practice. In particular, grain and rice left in the fields following mechanized harvesting has significantly increased the winter food supply for cowbirds. Cowbirds parasitize a range of songbird species by laying their eggs in the nests of these species. The parasitized hosts often raise the resulting offspring as their own. The true offspring may starve while the larger cowbird offspring monopolize the food brought by the parents. Because cowbirds don't have to rear their own young, female cowbirds can produce a lot of eggs over the course of the breeding season, and parasitize a lot of nests. As many neotropical migrant species only produce one brood of young in any breeding season, they are particularly susceptible to brood parasitism by cowbirds.

The success of cowbirds has been considerably aided by the fragmentation of forests into small patches. Because brown-headed cowbirds prefer to feed in open agricultural areas, they tend to penetrate only several hundred yards into the forest when searching for host nests to parasitize. Nests at the center of woods that are larger than a kilometer in diameter are relatively safe from parasitism. In a highly fragmented landscape, however, most woodlots will be small and without any substantial "core" area. Scott K. Robinson of the Illinois Natural History Survey and his colleagues have examined the nesting success of woodland birds throughout the Midwestern United States. They found that in a tract of wood near Shelbyville, Illinois, for example, 80 percent of all clutches laid were lost to predators, and that 76 percent of the survivors were parasitized by cowbirds. Their work shows that the rate of cowbird parasitism increases rapidly as forest is lost. A wood thrush population that would be mostly unmolested by cowbirds in a heavily forested area will have more than 70 percent of its nests parasitized once the percentage of land covered by forest falls below 40 percent.

Birds nesting in smaller woodlots are not only more susceptible to cowbirds, but also to nest predation by racoons, blue jays, and crows. The higher rate of predation and parasitism around the perimeter of forests is termed an "edge" effect. Forest edges are brighter and warmer, as well as drier and windier, than the forest interior, and they shelter more shrubs, vines, and weeds. The edge climate and vegetation usually extend only 10 to 30 meters inside the forest, depending on whether the edge has a northerly or southerly exposure. But David Wilcove, while at Princeton University, and others have shown by observing artificial nests placed at various distances from the edge that the edge-related increase in predation may extend from 300 to 600 meters into the forest. The nesting success of songbirds is considerably lower within 500 meters of a forest edge than in the interior, mainly because many nest predators—blue jays, American crows, common grackles, chipmunks, short-tailed weasels, and raccoons— as well as brood parasites, such as the cowbird, prefer the forest edges.

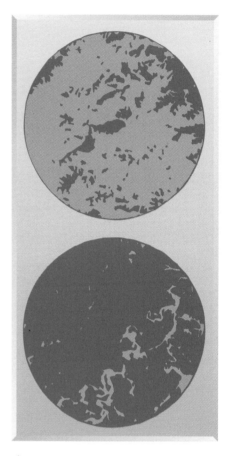

▲ These computer-interpreted maps show two landscapes in Missouri that were analyzed by Scott Robinson in his study of cowbird parasitism. Cowbirds were a much greater threat in the landscape shown at the top, where the proportion of forested land (green) is low.

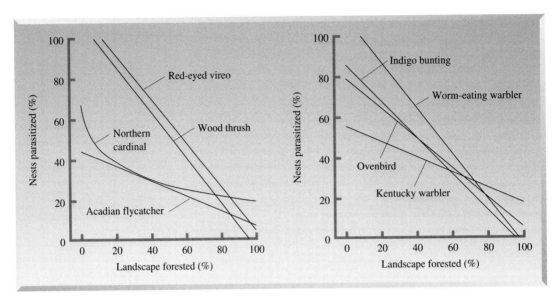

▲ Robinson's study demonstrated that, for many species of songbird, the percentge of nests parasitized rose rapidly as the percentage of forest cover dropped.

The disappearance of songbirds from many North American forests emphasizes that characteristics of the surrounding mosaic are often important in determining the viability of populations living in patches. If farmland attracts cowbirds to the area near a woodlot, brood parasitism will be high; it might be low, however, if the surrounding landscape has few cowbirds. Furthermore, the disappearance of the songbirds sharply illustrates that the degree to which edge effects extend into the remaining patches of habitats often determines the true size of the remaining forest fragments. Ironically, much of wildlife management in the early part of the century was aimed at creating edge for game species such as white-tailed deer and American woodcock, which do well in edge habitats. However, it is increasingly clear that the forest edge is a dangerous place for other members of woodland flora and fauna. Moreover, the presence of a 500-meter edge significantly reduces the effective size of any surviving fragment of woodland habitat.

Predicting the Consequences of Fragmentation

Conservation biologists face a frustrating dilemma when trying to monitor how habitat fragmentation alters ecosystem function. Habitat fragmentation is not a unitary process: it consists of a number of different mechanisms, of which the most important are the loss of total area from the habitat and its fragmentation into a system of smaller patches. Biologists often have trouble distinguishing which of these two processes is responsible for any one alteration in the ecosystem. Moreover, the full effects of fragmentation are often not seen for several decades.

Nevertheless, a number of projects have been set up to monitor the long-term effects of habitat fragmentation. Notable among them is the Biological Dynamics of Forest Fragments project in the rainforest of Brazil near Manaus. Initiated in 1979 (as the Minimum Critical Size of Ecosystems Project), this study was begun specifically to investigate the relationship between the size of a forest fragment and the stability and diversity of the species within it. The project took advantage of a Brazilian law that required 50 percent of land used in development projects in the Amazon basin to be left as forest. Through a series of agreements with local cattle ranch managers, the project researchers set up a series of forest fragments paired with similar sites in undisturbed forest. The project has now run for over 15 years and continues to produce interesting results. For example, the study indicates that a break of as little as 80 meters is sufficient to act as a barrier to the movements of many insects, some mammals, and most birds that live in the low-lying vegetation of the understory. One consequence is that far fewer flowers are visited by bees in patches that are only 100 meters away from continuous forest. Plants that miss these visits may never be pollinated or produce viable seeds. Similarly, dung beetle numbers dropped in the isolated 1- and 10-hectare fragments once the fragments had been isolated for two to six years. Dung is taking longer to decompose in these paatches; presumably nutrients are not being returned as quickly to the soil. Will the growth of vegetation suffer as a result?

To understand the often complex data produced by these large-scale experiments, ecologists have developed a number of mathematical and computer models. These are designed to examine the rates at which species are lost as hypothetical natural habitats are reduced and fragmented. The

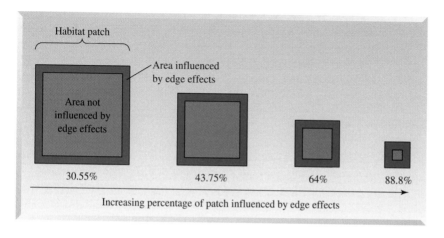

◀ As the size of a habitat shrinks, so does the percentage of area free from edge effects.

simplest models assume that a community consists of a group of species, each of which has a different minimum area requirement. In most natural communities, only a few species have large area requirements; those for the majority will be relatively modest or even small. If we assume that the distribution of minimum area requirements follows a bell-shaped curve when plotted along a logarithmic x-axis, (called a log-normal distribution), then we will probably have a pretty good approximation to the situation in most natural communities. We can now take a hypothetical chain saw to the habitat and examine what happens to the diversity of the community as we randomly fragment the habitat into many smaller patches.

Initially let us consider how fragmentation proceeds as the original habitat shrinks: if areas of habitat are cleared in a random sequence, then initially contiguous habitat usually breaks into several patches when around 40 percent of the original habitat is lost. At this point the size of the largest fragment drops rapidly. Whereas shortly before, a single patch still filled between 40 and 50 percent of the original area, there are now suddenly many small patches, each taking up approximately 2 percent of the original area. As fragmentation proceeds, the average size of the remaining patches decreases while the distance between neighboring patches widens. That distance increases sharply once less than 20 percent of the original habitat remains. In real cases, however, the course of fragmentation doesn't follow quite this pattern, because the sequence of areas cleared is not random. Instead, sites of clearing tend to be aggregated around roads passing through the habitat, as has happened around the trans-Amazon highway in the state of Rondonia in Brazil. The result is that more habitat can be removed before the original contiguous habitat breaks into several patches.

The significant question is what happens to the species inhabiting the community. Initially, we see a modest loss of species for quite a substantial reduction in area. However, as habitat loss proceeds and the net area is reduced to many fragments, each much smaller than the original habitat, then we begin to see a rapid climb in the rate at which species are lost.

The great value of these computer models is that we can modify this initial basic framework to specify explicitly those characteristics of plant and animal species that we are interested in exploring. For example, we can assume that some species are better at finding new fragments of habitat than others. As we might expect, the species that leave their home territory only reluctantly go extinct at faster rates than those better able to disperse and find new homes.

More detailed models have examined rates at which species are lost when the competition between species for resources is taken into account. Several important points emerge from these studies: the key message is that it takes much longer for species to become extinct than it does for their habitat to be reduced. David Tilman and his colleagues suggest that although habitat destruction may lead to the immediate loss of some

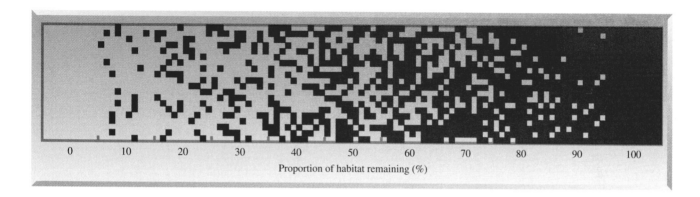

Proportion of habitat remaining (%)

species, the majority of extinctions will occur a long time after the initial loss of habitat. They call this phenomenon an "extinction debt;" a consequence is that although the costs of habitat destruction may appear light at first, the full costs will steadily increase and be paid over a very long time.

Tilman's analysis also suggests that even before many species are lost, species may move drastically up or down in the rankings for relative abundance. In the intact habitat, the most common species are those that compete most successfully for resources. When a proportion of the habitat is destroyed, the common species are rapidly replaced by species with less ability to compete, but greater ability to disperse into unoccupied patches of habitat. In some ways, something similar occurred when the perennial grass species in the Sahel regions of West Africa were overgrazed around water holes and replaced by less palatable, shallow-rooted annual species.

Computer models are wonderful for revealing trends that might otherwise be missed in the jungle of (highly incomplete) data, but because they cannot completely capture the complexity of the real world, their results are always somewhat suspect until confirmed by studies in the field. Unfortunately, only a few long-term studies have monitored how the abundances of all the resident plant and animal species are changing in habitats that are undergoing destruction and fragmentation. The data sets described above for birds in North American forests, and those compiled for a variety of species in the Brazilian rainforest, provide ominous support for many of the patterns described in the theoretical models and computer simulations. For example, in his study of cowbird parasitism, Robinson found that once the extent of forest cover drops below 60 percent, the habitat becomes highly fragmented and rates of cowbird parasitism increase rapidly.

Computer models have missed one recurring tendency that is apparent in many field studies of fragmentation in the tropics: that is that many large vertebrate species tend to go extinct faster than we would expect from their area requirements. Although not predicted by computer models, this effect is obvious to anyone who has ever visited a pristine rainforest. In the

▲ This computer-generated landscape illustrates how the destruction of an increasing proportion of an originally continuous landscape produces fragments of ever smaller size. The dark area represents the original habitat and the light area the extent of habitat loss.

tropics almost anything that is large enough to eat is quickly eradicated by human hunters within a couple hours' walk of encroaching roads and agricultural developments.

Although the theory of island biogeography provides some useful insights into the impact of habitat destruction, we have to always bear in mind that habitat fragments are not real islands. Rather, they are part of a landscape mosaic with permeable boundaries, subject to processes originating both within their borders and without. In some cases, a population will "spill over" from a patch of ideal habitat into less suitable habitat in the surrounding mosaic. Similarly, predators that live in the surrounding mosaic may find the patch of habitat a useful place to forage. Moreover, fragmentation is dynamic: once cleared, and if not put to permanent other use, land in the surrounding mosaic may revert to its original condition, passing through a succession of stages, each characterized by a unique mix of species, on its way back to its starting place. In the minimum viable ecosystem project in the rainforest near Manaus, Brazil, species from the original patches were able to recolonize the surrounding landscape with surprising rapidity. But to balance this note of optimism, it's only fair to emphasize again that many populations will experience time lags in their response to fragmentation. An examination of the species that are present immediately after fragmentation may well give a false impression of the species that will ultimately inhabit any patch.

Ultimately, the influence of habitat loss upon the decline of any particular species is very tricky to diagnose. As Graeme Caughley has pointed out, "There are examples of researchers assuming that the endangered remnants of a once widespread species have settled upon the most favorable habitat left and that their present life cycles and feeding habits represent normality. A safer preliminary hypothesis would conjecture that the species ends up, not in the habitat most favorable to it, but in the habitat least favorable to the agent of decline." Caughley's point re-echoes the point made earlier: humans are not only unique in their ability to destroy natural ecosystems, they are unique in having the opportunity to decide which areas should be set aside to conserve other species. Unfortunately, it seems likely that the need for agricultural land will ensure the continued destruction of natural habitats for the immediate future. In order to give the remaining biodiversity its best chance of survival, conservation biologists are developing some of the predictive techniques described in this chapter to decide where we should put nature reserves, and how their size, shape, and management affect our ability to conserve ecosystems and their species.

The portrait of habitat destruction that I have painted in this chapter is profoundly disturbing to me and other conservation biologists. We suspect that massive extinctions should be taking place as the logical consequence. The famous disappearance of the dinosaurs 65 million years ago tells us

◀ Most large species, such as this caiman, are quickly lost when humans colonize natural habitats. They are either poached for their skins or eaten as a source of protein.

that catastrophic extinctions are not unknown in earth's history. Could species be disappearing now at rates that rival, even exceed, those during that well-known episode? How can we tell how many species are going extinct if we can't even count how many inhabit our world to begin with? As the next chapter shows, biologists are finding answers to these questions by exploiting some of the same insights from the study of islands that have proved so useful in understanding fragmentation.

Chapter 3

*T*he Mathematics of Extinction

For the first Earth Day, held on April 20, 1970, a cemetery for extinct species was erected on the main lawn at New York's Bronx Zoo. A gravestone marked the extinction of each species, with a row of gravestones for each fifty-year period of the last five hundred years. Each year on Earth Day the extinct species cemetery is re-created, and each year more species are added to the rows of memorials. The scene creates a haunting sense of loss that echoes that evoked by the mass military cemeteries in Flanders and Normandy. In retrospect, we may eventually appreciate that the loss of these species marks a more pronounced turning point in the earth's history than either of the two world wars that have dominated this century.

The exhibit at the Bronx Zoo sharply illustrates how many species have already been lost and how quickly the rate at which they are disappearing is rising. The growing number of extinctions is the most irreversible of the impacts produced on the earth by a rapidly expanding human population. However, human births are

◄ A model of the extinct dodo *Raphus cucullatus*, based on the few remaining skeletons, a foot cast, and a stuffed head. This flightless member of the pigeon family was hunted to extinction only a century after its discovery on the island of Mauritius in 1592.

▶ At the Earth Day Cemetary in the Bronx Zoo, a section on endangered species warns of possible future extinctions.

translated into the extinctions of other species through a complex network of processes. Population biologists have constructed mathematical models to assess the convoluted processes that have contributed to extinction. By probing the mathematics of extinction, the models have revealed a number of interesting patterns. Crucially, they have demonstrated that small, isolated populations almost inevitably dwindle in size and disappear.

Extinction Rates—Past and Present

The extinctions of bird and mammal species are the most completely documented in the historical record. Since 1600, the populations of some 113 species of birds and 83 species of mammals are known to have completely disappeared. About three-quarters of these species lived on oceanic islands. In many cases the cause of the extinction remains unknown—often because the only record of the species we have is a museum specimen collected on a scientific expedition. After the expedition left the island, the species was never seen again, and it is not known whether, for example, habitat loss, overhunting, or the invasion of new pathogens led to its demise. The historical record further suggests that the rate of extinctions has increased, by at least a factor of 4 since 1600, to produce the current extinction rates of around one-half percent of extant birds per century and 1 percent of extant mammals. Indeed, these figures are likely to underestimate the real rates of extinction for these groups, since undiscovered bird and mammal species living in tropical forests will go extinct before they are found and described by scientists.

Why did the majority of bird and mammal species subject to recent extinctions live on oceanic islands? The fossil records tell us that the number of extinctions often climbed steeply after the arrival of humans on previously isolated oceanic islands or the smaller isolated continents. Typical are the patterns of extinction observed in three of the last areas on earth to be colonized by our species: Hawaii, New Zealand, and Madagascar.

According to the fossil records, around 50 percent of the endemic birds in Hawaii went extinct following human colonization 1500 years ago. In Madagascar, a considerable number of species—*Aepyornis* (the elephant bird), as many as a dozen lemur species, giant land tortoises, and a species of hippopotamus—all went extinct after humans arrived around A.D. 500. In New Zealand, 12 to 13 species of moas (large flightless birds) disappeared following the arrival of the Maoris around A.D. 1000. Recent studies of bones from small birds on oceanic islands in eastern Polynesia suggest that more than 2000 species of birds may have been extirpated in the Pacific since humans colonized these islands within the last several thousand years. Had they survived, these species would have raised the total number of known bird species from 8600 to 10,600—an increase of nearly 20 percent!

An island's human arrivals posed a threat to its older inhabitants in two ways. Cut marks, made by tools, found on ancient bones are a sign that

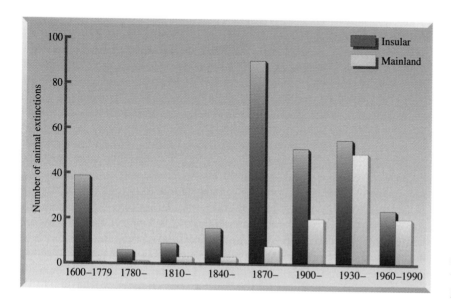

◀ Before 1930, island extinctions considerably outnumbered mainland extinctions; in the years since, mainland extinctions have just about caught up.

▶ Delalande's cuckoo, *Coua delalandei,* is known only from 13 specimens, the last collected in 1834. This extinct bird of Madagascar fed on large snails whose shells it broke against a stone. The bird's lowland rain-forest habitat has long since disappeared.

humans exploited large animals for food. But humans put smaller organisms at risk, too, by introducing domestic livestock, such as goats, dogs, cats, and particularly rats. The voracious grazing of goats could turn a landscape of succulent grasses and flowering trees into a tract of thorny bushes that neither goats nor native species can easily feed on. While goats devastated the island flora, cats, rats, and dogs hunted the local fauna as a source of food. Because island species have usually evolved in the absence of predators and herbivores, they have few natural defenses against these threats and are easily preyed upon by humans and their domesticated companions. Even the flying ability of birds does not help them. Many island bird species are flightless or nest on the ground, where cats and dogs can easily prey on the eggs. And those species of bird that nested in trees were vulnerable to the incursions of tree-climbing rats and cats.

As the preceding example illustrates, a species' own behavior or other characteristics can make it more vulnerable to a new predator, loss of habitat, or other threat to its existence. After examining bird species from the tropical Americas, John Terborgh and Blair Winter have suggested that certain species are especially prone to extinction. In particular, species that do not readily disperse to new sites, and species with chronically small populations, are highly prone to extinction. Similarly, local endemic species that are large when compared to closely related species or that feed higher in the food chain, such as many predators, tend also to be at risk. Large animals and predators are made vulnerable by their small numbers within any area: each animal requires a relatively large territory to supply its food, so a crowded population would risk starvation. Finally, migratory species and those that nest in large social aggregations are particularly prone to extinction.

The number of large species that have vanished quite recently is especially striking. David Burney has discovered by inspecting the recent fossil record that only 12,000 years ago the landscape of the southern United States rivaled that of East Africa today in its diversity of large mammals. The fauna of that time and place included woolly mammoths, mastadons, and giant ground sloths as well as species that are extinct in the Americas, but survive elsewhere, such as cheetahs, lions, zebras, and tapirs. This wildlife spectacle ended during the Pleistocene era about 11,000 years ago as human populations began to expand—North America, and indeed most other continents, lost more than half its mammal species and most of the largest ones.

▼ Bones from an extinct flightless ibis, *Apteribis glenos*, of Hawaii are the only signs we have that it once existed.

Predicting Future Extinctions

While island species are clearly vulnerable to extinctions, evidence is accumulating to suggest that rates of extinction are increasing in mainland species as well. A recent survey of North American fish species, for example, found that these species are becoming extinct at exponentially increasing rates. Nearly half the species lost this century have disappeared in the last ten years. Furthermore, in the United States, there were nearly twice as many fish species officially classified as endangered (350 species) in 1990 as there had been ten years previously; 3 genera, 27 species, and 13 subspecies have already become extinct during this century.

Many of the species (73 percent) were lost when their habitat was so altered that they could no longer survive. An example is the fate that befell the thicktail chub, a fish of California's Central valley. This species, which lived in the stagnant waters of streams, sloughs, and ditches, seems to have been common until the end of the last century: archaeologists excavating Native American middens have found it to be the third most abundant food in the native diet, and during the last century the fish could be purchased in the market in San Francisco. Then, in the twentieth century, the backwaters of dams flooded many of its native streams; other streams disappeared when their waters were diverted to the irrigation systems of the Central valley. By the 1950s the last of the chub had disappeared. Similar mammoth recasting of the landscape continues. By removing meanders,

▶ These Borax chub, from Borax lake in the state of Oregon, have been classified as endangered. Because lakes, ponds, and streams are usually isolated from one another, each is like an "island" of habitat in which populations may evolve into new species. Thus many species of fish live only in a single pond or stream; if its waters disappear or become polluted, these fish become extinct.

for example, the Army Corps of Engineers has shortened the Mississippi river by 150 miles and reduced the area covered by wetlands by more than 17 million acres in the states of Illinois, Missouri, and Iowa (an acreage amounting to 85 percent of the wetland areas across these states). It would be too much to hope that the river's fish populations will emerge unscathed.

Most North American fish species that have gone extinct have faced more than one threat, and the thicktail chub was no exception. The coup de grace for the chub came from a seemingly innocuous source, the introduction of the Hitch minnow into the chub's remaining habitat. The two species are similar enough that they will breed with one another, producing hybrid offspring. Hybridization is considered undesirable for most species, since, if any offspring are produced, they are usually sterile and they have caused their parents to lose a breeding opportunity. Hybrids between the thicktail chub and the Hitch minnow were observed as early as the 1920s, and, although these hybrids were not necessarily sterile, they spelled the chub's doom as a pure-bred species.

Hybridization and habitat alteration do not exhaust the threats to North American fish. Of the species that became extinct in the past hundred years, 30 percent were endangered by pollution and hybridization and 68 percent by the introduction of competing species or predators.

Similar alarming stories of widespread declines are beginning to emerge for reptiles, amphibians, and fungi throughout the world. Because fewer people are actively studying these groups, it has proved less easy to verify that the declines are real, although a series of detailed surveys from locations throughout Europe and North and Central America suggest that they are. Unfortunately, as we move away from birds and mammals, our knowledge of the distribution and abundance of species from different taxa declines precipitously. Thus although the International Union for Conservation of Nature has officially classified only a very low percentage (less than 2 percent) of mollusk, insect, and reptile species as endangered, this low percentage may simply reflect ignorance of the status of many species, rather than optimism for their continued well-being. To obtain any estimate of the true rates of loss for these groups of species, we must turn to indirect methods.

The major factor driving species to extinction at the present is the destruction of their natural habitats, the forests, savannas, and other ecosystems in which they live. But it is the destruction of the world's tropical forests that is driving species extinct at the fastest rates. How many species are lost will depend in some way on how much habitat disappears. Is there some means through which we can make a connection, in hard numbers, between hectares lost and species made extinct? In fact there is, and the key to the method is the species-area curves described in Chapter 2. These curves tell us how many species will be found in a region of particular size.

▲ This nectar-feeding species of Hawaiian honeycreeper, *Hemignathus lanaiensis*, was last reported seen in 1894. Hawaii's 22 species of honeycreeper (of which 9 are now extinct) evolved from a single ancestor that migrated from America. Three-quarters of the bird's forest habitat has disappeared, much of it overgrazed by feral cattle introduced by Europeans. Honeycreepers also have no immunity to the avian malaria brought to the islands by imported birds.

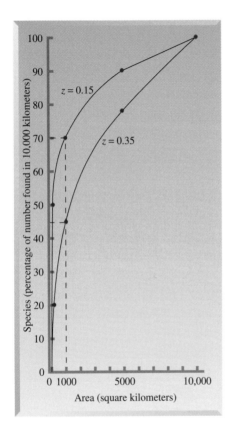

▲ Two species-area curves of different slope give the percentage of species lost from a region as the area of habitat contracts, assuming that 100 percent of the species are found in a habitat of size 10,000 square kilometers. Using curves like these, it is possible to predict the number of extinctions caused by habitat loss. On an island (for which the slope $z = 0.35$), the loss of a defined amount of habitat will cause more extinctions than the same habitat loss within a continent (for which the slope $z = 0.15$).

Interpreted cautiously, they can tell us what proportion of species will go extinct as that size shrinks.

Walter Reid of the World Resources Institute has illustrated how we can use species-area curves to estimate the present rate at which tropical forest species are becoming extinct. Three pieces of information are needed to apply effectively the technique Reid (as well as others) uses: the current area covered by forest (or other habitat), the rate at which habitat is being lost, and the slope of the species-area curve, which will tell us how quickly species numbers fall with declining area. Reid chose to look at the loss of "closed" tropical forest, the dense forest that supports the great majority of tropical species. The most recent estimate, arrived at in the 1980s, for the total area covered by closed tropical forest gave a figure of around 1166 million hectares—about 80 percent of the forest's original extent. Approximately 15 to 20 percent of this forest area has already been disturbed by logging and other activities. A number of different organizations have made estimates of deforestation rates for tropical forests: these range from 14.2 million hectares of closed forest annually, according to independent consultant Norman Myers; to 16.8 million hectares of closed and open forest, according to the United Nation's Food and Agriculture Organization; to 20.4 million hectares of tropical forest, according to the World Resources Institute. The wide spread in these figures suggest they are uncertain, but coming up with these figures has been complicated in part by the attempts of various governments and the international timber trade to disguise the actual logging rates. The difficulty is compounded by the fact that some countries are reducing their rates of logging, while others are under pressure to log forests even faster in order to reduce government debts. Walter Reid has taken this uncertainty into account by calculating extinction rates twice: once for a global deforestation rate of 10 million hectares per year and once for a rate of 15 million hectares per year. The lower rate is close to the latest best estimate of levels of closed tropical forest deforestation, while the upper rate is closer to those quoted above, and acknowledges that, even if the lower rate is correct now, the present rate of tropical deforestation is likely to increase by 50 percent.

Species-area curves represent the relationship between the number of species found in an area and its size. Although a number of equations can be used to describe this relationship, the simplest and most widely used equation assumes that the number of species increases as a simple power function of area:

$$S = cA^z$$

where S is the number of species, A is the area, and c and z are constants. Part of the convenience of this formula lies in the fact that when area and number of species are converted to logarithms, the curve can be plotted as a straight line whose slope is z:

$$\text{Log } S = \log c + z \log A$$

Species-area curves of this type have been fitted to a wide range of sets of ecological data both for real oceanic islands and for habitat fragments, such as woodlots, that act as surrogate islands. They characteristically explain greater than 50 percent of the variation in the number of species observed in each patch. The slope of the line, z, invariably lies in the range between 0.15 and 0.35. The slopes derived from studies of habitat fragments tend to lie toward the low end of this range. The species-area relationship roughly suggests that destroying 90 percent of a habitat will lead to the extinction of 50 percent of the species living exclusively in that habitat; destroying 99 percent of the habitat will lead to the extinction of 75 percent of the species, and so on.

From the relationship described in the species-area curves, we can predict the percentage decline of closed tropical forest species, if we assume that forests will continue to be destroyed at their present rates for the next 25 to 50 years. Here we should recognize one very important caveat: the technique calculates the proportion of species that will *eventually* decline to extinction. These species may take considerably longer to disappear than the 25 to 50 years for which habitat loss has been monitored. As I mentioned in Chapter 2, habitat destruction creates an extinction debt that may not be paid instantaneously. Indeed, it can take at least a century for the full effects of isolation to take effect, as ecologists have observed in detailed studies of bird and plant species on Barro Colorado Island in Lake Gatun in the Panama Canal. It only took a few years to create the island when the Panama Canal was flooded in 1914, but many species that later disappeared persisted for decades after the island's isolation. In fact, extinctions continue to take place as a result of the island forest's isolation from the forest on either side of the lake.

If we return to our tropical forest analysis, we find that, at global deforestation rates of 10 million hectares per year, species will go extinct at the rate of 1 to 2 percent per decade in Africa, and 2 to 5 percent per decade in Asia. On a global scale, the technique predicts that we will commit 4 to 8 percent of the world's tropical forest species to extinction in the next 25 years and as many as 9 to 19 percent over the next 50 years. If there are roughly 5 million species on earth, and 50 to 90 percent live in tropical forests, this would amount to between 4000 and 14,000 species per year, or 10 to 38 per day. In a world containing 5 million species, this suggests that the average "life span" of any species is between 200 and 500 years. Although these rates are an approximation, they are based on the best

Year	Region	Mid Scenario (10 million ha/yr)	High Scenario (15 million ha/yr)
2015	Africa	3–6	4–9
	Asia	5–11	8–18
	Latin America	4–8	6–13
	All tropics	4–8	6–14
2040	Africa	6–13	10–21
	Asia	12–26	28–53
	Latin America	8–18	15–32
	All tropics	9–19	17–35

Predicted Percentage Decline in Tropical Forest Species Between 1990 and Year Indicated

available estimates of forest cover and habitat loss, so the actual rate is probably well within this range.

The fossil record tells us that the majority of species that have existed on the earth have gone extinct. Extinctions are an inevitable part of the history of life on earth, and surely all species now living will be extinct someday. We might wonder if the current extinction rates truly present a problem, and whether these rates are unusually high or just the norm. How do our estimates of current rates of extinction compare with those in the fossil record?

Extinction in the Fossil Record

The extinction of an established species is almost as common an event in the fossil record as the appearance of a new one. Indeed, the number of species of animals and plants alive today constitutes around 2 to 4 percent of the total there have ever been. In the ancient past there have been at least five examples of mass extinctions, when extinction rates rose to very high levels. The most serious of these came at the end of the Permian period, 250 million years ago. Because it can be difficult to identify individual species from fossil material, extinction rates are only available for families of organisms. Nevertheless, it is estimated that at the end of the Permian period perhaps 60 percent of the animal families then living became extinct (63 percent of continental organisms and 48 percent of marine organisms). Attempts have been made to interpolate the rate of family extinction to ob-

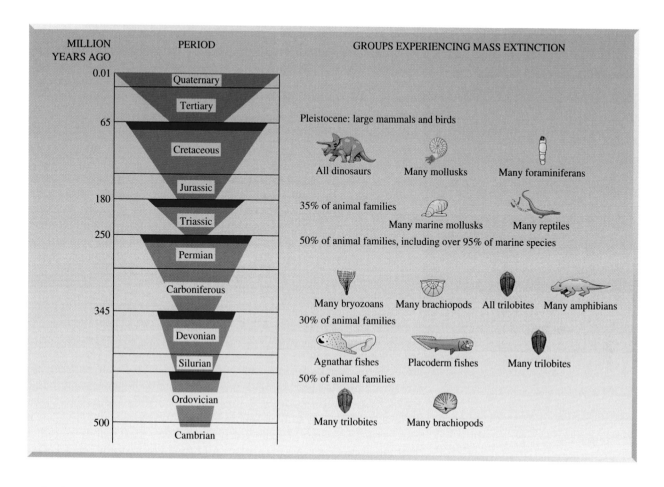

| MILLION YEARS AGO | PERIOD | GROUPS EXPERIENCING MASS EXTINCTION |

Pleistocene: large mammals and birds

All dinosaurs   Many mollusks   Many foraminiferans

35% of animal families

Many marine mollusks   Many reptiles

50% of animal families, including over 95% of marine species

Many bryozoans   Many brachiopods   All trilobites   Many amphibians

30% of animal families

Agnathar fishes   Placoderm fishes   Many trilobites

50% of animal families

Many trilobites   Many brachiopods

tain the levels of species extinction; they have yielded estimates ranging from 77 to 96 percent for the percentage of marine animal species that became extinct at the end of the Permian period. Although the percentages seem enormous, these extinctions took place over a period of around 5 million years, so that the percentages are much lower over smaller time scales.

Exactly what caused the Permian extinction is unknown, but several strands of evidence suggest that climatic cooling played a large role. The geographical ranges of several groups of organisms contracted toward the equator before the organisms died out, as though they were retreating to the last remaining habitats of warm temperature. In addition, oceanic cooling seems to be the most plausible explanation for why two types of reef builders disappeared for a few million years after the close of the Permian. Why the climate should have cooled remains an unanswered question.

Another four extinction events are also usually classed as mass extinctions, although each of these, too, took place over more than a million

▲ Although there have been many lesser episodes of extinction, the five major ones took place at the end of the Ordovician, Devonian, Permian, Triassic, and Cretaceous periods.

years. The most recent mass extinction occurred 65 million years ago at the boundary between the Cretaceous and Tertiary periods. Each of these large extinctions is thought to have eliminated at least half the animal species living. Like the Permian extinction, each of these large-scale extinction events seems to have taken place during a period of climate change.

It is possible to compute average rates of extinction from the fossil record. David Raup of the University of Chicago has estimated that on average about 9 percent of species become extinct during any million-year period, a rate that suggests the average lifetime of a species is around 10 million years. This estimate of background extinction rate is probably low, by at least a factor of 10, because of a sampling bias (namely, paleontologists will not usually find more than one set of fossils, from one moment in time, for a species that did not spread beyond a small area, no matter how long the species persisted). The figure of 10 million years means that in a biosphere containing 5 million living species, about one species is lost every two years. By definition, this estimate is very sensitive to estimates of the numbers of species that exist. Since our estimate of the number of species may be wrong by a factor of between 2 and 10, the average background extinction rate may be as high as one to five species per year.

If we briefly return to the comparatively well studied birds and mammals, we find that extinctions this century have indeed been running at the rate of around one per year, but these groups only contain around 10,000 and 5000 species respectively. If this rate were maintained in the more numerous but less well studied invertebrates, then the average "species lifespan" would be around 10,000 years, two to three orders of magnitude less than the average species lifespan seen in the fossil record. Yet it is two to three times longer than estimates obtained from considering the rate of habitat loss. What causes this difference? In part the longer estimate obtained by extrapolating from the known extinctions reflects an underestimate of the true rate of extinctions. Many species on unvisited oceanic islands go extinct without anyone recording their passing, as do many species in tropical forests. In contrast, extinction rates based on rates of habitat loss account for unrecorded species fairly reliably as long as our estimate of the total number of species is correct.

A number of estimates in addition to Reid's have been made of current rates of extinction for all known species; the highest of these give a figure between 10 and 25 thousand species per year, or one to three species every hour. These estimates suggest that species are going extinct more than ten thousand times faster than at any previous time. Around half the world's existing fauna and flora evolved in the last 50 to 100 million years, and yet if we assume that habitat destruction will continue at its present rate, then a similar proportion of the species now living will be lost in only the next

50 to 100 years. These figures suggest that present extinction rates exceed speciation rates by a factor of around one million. If true, the result can only be a major extinction spasm that far exceeds anything seen in the fossil record.

Stuart Pimm, of the University of Knoxville in Tennessee, has recently explained in very simple terms why we need to worry about the current rates of animal and plant extinctions. Speaking to a panel of U.S. congressmen who were trying to understand why the current extinction spasm is important if all species are doomed eventually, he said, "Everyone on this panel will die some time in the next ten to fifty years—it will be sad, but it won't be a tragedy. If we all die in the next week, then it's a tragedy."

Quantifying the Risk of Extinction

Simply being able to give a figure for the number of future extinctions is not enough to satisfy the scientist's curiosity. Conservation biologists want to understand the ecological processes that contribute to extinction, and, more practically, they want to know how large a population need be in order to be considered "safe."

A declining population is one in which, obviously, individuals are dying, but it is also one in which not enough new individuals are being born to replace the ones that disappear. Clearly, some force or forces must be acting to increase the death rate or decrease the birth rate. Ecologists want to know how these rates will respond to loss of habitat, the introduction of predators, and other human-induced change. The problem is that, even in the most pristine environments, populations are constantly rising and falling as they respond to natural forces—the freak flood, the periodic drought, the increase in number of a predator, the drop in number of a food plant, and so on. Some of these fluctuations could be the result of a long-term trend: the climate may be changing, or natural selection may be improving a population's ability to use its food resources. Other fluctuations are the response to normal variations in climate or resources. These natural forces could be reinforcing the human-induced forces or counteracting them. The goal is to understand how populations rise or fall in response to these forces.

An important next step toward this goal is to develop an understanding of the mathematics of population growth and decline. Ecologists have explored ways of expressing environmental change numerically and have developed equations that describe how birth and death rates rise and fall in response. These equations are the backbone of mathematical models that allow ecologists to predict the fate of a given population, whether it will prosper, barely eke out an existence, or fade away entirely.

Initially, let us consider the factors that led to the extinction in the wild of the black-footed ferret, *Mustella nigripes*. In February 1987, the last known black-footed ferret was brought into captivity. The number of individuals in the last surviving population had been steadily declining since the rediscovery of the species in September 1981 at Meeteetse in Wyoming. Ferrets originally began dying off throughout their range when the prairie dog colonies, which form the basis of the ferret's main food supply, were poisoned by farmers. Epidemics of canine distemper then wiped out most of the rest of the population before the decision was made to bring the remaining individuals into captivity.

Fortunately, the species has bred well in captivity and in 1992 some 50 individuals were reintroduced back into the wild. At first the reintroduction was not thought a major success, as none of the 50 individuals could be accounted for a month after their release. However, it was later discovered that at least two released females had secured mates and produced litters. Further reintroductions are now being made to enlarge the newly established free-living population.

The black-footed ferret's extinction and the success of any reintroduction program are dependent on a mixture of forces. Some are predictable (or deterministic), such as the age at which a female ferret first reproduces, and others are variable (or stochastic), such as the severity of the winter or the ratio of male to female kittens in a particular litter. Ideally, any mathematical model we build for the black-footed ferret should consider both deterministic and stochastic forces. However, as stochastic models require more complex mathematics than deterministic models, let us begin by considering simple deterministic models for population growth and decline, explained in more detail in the box on the facing page.

Consider the graph on the next page, which gives a record of how the numbers in a black-footed ferret colony fluctuate in the years from 1982 through 1985. At first glance, the population doesn't seem to show any consistent trend: the numbers swing up and down every year. But that swing hides a fairly steady pattern. Unlike humans, ferrets do not reproduce at just any time of year. Rather, a female ferret bears her litter predictably in the month of July, when she has more time during the long days to hunt for the extra food she needs to feed her young. The population shoots up during that month, then gradually declines the rest of the year as the weaker newborns die and older ferrets fall victim to predators or perish during the harsh winter months. To make trends in population growth visible, the population must be counted at the same time every year.

Suppose we select the month of December in the years 1982 through 1984. We record 11 individuals in December 1982, 15 individuals in December 1983, and 17 individuals in December 1984. The rate of popula-

▲ Several black-footed ferrets in captivity share a nest. In the wild a family of the ferrets would nest in a prairie dog burrow.

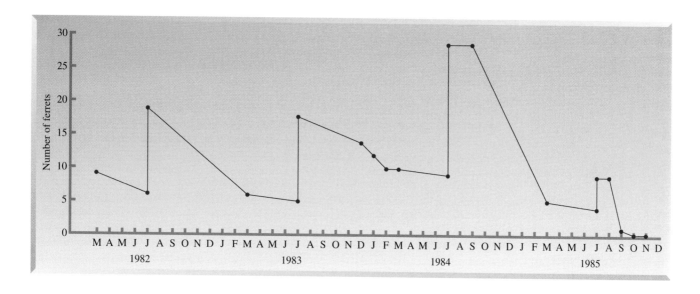

▲ This graph records the changing numbers of the last known wild population of the black-footed ferret, in the final years of its existence.

tion increase is the ratio of the numbers in one year to that in the preceding year: 15/11 for 1983 and 17/15 for 1984, altogether an average population growth rate of 1.25 per year for the two years combined.

The rate of population increase is a fairly simple consequence of the death rate and birth rate, as explained in the box on the next page. A deterministic mathematical model would take values for the proportion of individuals who survive every year and for the average number of offspring born to each female, and come up with an annual rate of change, which I have called R in the box. From this rate, the model could calculate the future course of a population's rise or decline.

Deterministic models assume that the birth and death rates remain constant. In the simplest deterministic case, populations will always increase in size if the birth rate exceeds the death rate. That seems to be what is happening in the black-footed ferret population, at least at first. If we take the value of 1.25 computed for the growth rate and calculate the future course of population growth, we find that it should increase to 21 individuals in 1985, 26 individuals in 1986, and so on. But a glance at the graph shows that something entirely different took place: in 1985 the population suddenly plummeted to zero. The entire colony was wiped out by an outbreak of the virus that causes canine distemper.

Clearly, the deterministic model, while it seemed to work for a while, proved inadequate to explain, or predict, the real behavior of the population. To try to obtain more realistic results, conservation biologists have turned to more sophisticated models that recognize that variability is inevitable and that catastrophes and unforeseen events occasionally occur.

The Mathematics of Population Growth and Decline

The number of individuals in any population changes through time as death rates and birth rates fluctuate. In the simplest case, we can write an equation linking the size of the population, N, at successive times, t and $t + 1$, which are one year apart:

$$N_{t+1} = RN_t$$

Here R is the change in the size of the population in any one year. If R is greater than one, the population will increase in size; if R is less than one, the population will decline to extinction.

For most species the annual change in the population size is determined by the birth and death rates. For example, consider a small monogamous bird that breeds once a year from age 1 till the end of its life. The numbers of birds in successive years will be given by the equation.

$$N_{t+1} = sN_t + sFN_t/2$$

In this expression, we assume the birds have a constant annual survival, s (the proportion of the population that survives from one year to the next) and that each pair of breeding birds produces a constant number of chicks, F. To obtain the total population size, N, after a year-long interval, one adds the number of adults that survive between successive years, sN, and the numbers of surviving young produced by each female in the population, $sFN/2$. The expression can be rearranged to provide an expression for the annual rate of change, R:

$$R = s(1 + F/2)$$

The population will always decline to extinction when $s(1 + F/2)$ is less than one.

A growing population can survive with low rates of annual reproduction as long as the annual survival, s, is high. In contrast, should fewer individuals survive the year, then the annual fecundity, $F/2$ must rise to ensure that population growth, R, is greater than 1. It is relatively straightforward to include more details in this framework, such as the lower survival of immature birds or later ages at first reproduction.

Stochastic Models for Extinction and Persistence

Deterministic models of animal and plant populations simplify matters considerably by assuming a constant birth and death rate. However, even in populations living under controlled laboratory conditions, those rates vary continually from year to year. Some females will produce larger litters than other females; one year more individuals may die young from random accidents; another year more may survive to reproductive age. A population model that makes accurate predictions needs to capture this kind of variability.

In natural populations of animals and plants we recognize three types of variability:

1. *demographic stochasticity*, which arises from chance events in the birth and survival of discrete individuals.

2. *environmental stochasticity*, which is caused by changes in the weather, food resources, and other features of a population's habitat.

3. the variability resulting from *natural catastrophes*, such as floods, fires, and droughts, which occur at unpredictable intervals.

These different sorts of variability can be examined using stochastic models for population growth. In contrast to the deterministic models described above, which make a specific prediction that a population will number approximately X individuals after t years, we now calculate the probability of a population being a particular size. The calculation is complicated, and usually only special cases may be solved exactly. Nevertheless, it is a straightforward exercise to examine the properties of more detailed stochastic models using a computer. Unfortunately, the more details we include, the more sources of uncertainty we have to consider, and the wider in range our estimates of expected population size become.

Initially, let us consider the effects of demographic stochasticity on a simple population model. Demographic stochasticity represents the chance events that affect the births and deaths of individuals in a small population; in particular, it acknowledges the fact that births and deaths are events that affect discrete individuals. These individuals do not themselves have survival rates, which are properties of the population as a whole; instead, each individual has some constant probability of dying in any one time interval. Similarly, any female has a constant probability of producing an integer number of male and female offspring in any one time unit.

The effect of demographic stochasticity can readily be seen in our ferret example. If we extrapolate forward assuming the population grows at

▲ Studies by Joel Berger of the University of Nevada, Reno show that small populations of desert bighorn sheep almost inevitably decline to extinction.

our calculated rate of 1.25, we find that 17 ferrets alive in 1984 times 1.25 equals 21.25, but we can't have 0.25 of a ferret! So we round down and produce a slight reduction in the growth rate. The rounding can be either up or down, but the very fact that births and deaths come in integer values introduces variation into the growth rate.

The properties of stochastic models were first explored by Robert MacArthur and E. O. Wilson in the sixties and by Nira Richter-Dyn and Narenda Goel of the University of Rochester in the early seventies. When, borrowing from models originally developed to examine gambling in casinos, they factored in demographic stochasticity, they found that, in addition to the population growth rate, population size is now important in determining how long a population persists. In particular, it is possible, and often inevitable, for a small population with a positive growth rate to decline to extinction. The rounding down works against an endangered species of small population in the same way that gambling works against people who go to the casino without much money. If there is a minimum stake, poorly funded gamblers have fewer opportunities to bet and therefore fewer opportunities to bet again. The models thus suggest that there is a threshold population size, below which a population has a high probability of declining to extinction. A classic example of an extinction threshold was revealed by the long-term monitoring of bighorn sheep populations in the southwestern United States, which found that no population of fewer than 50 animals was able to persist for more than fifty years.

Small populations are more vulnerable to extinction, because in these populations the force of demographic stochastic variability is more strongly felt. Any populations can in some year experience a random cluster of misfortunes that carries off several extra individuals, while fewer females than expected become pregnant or they produce litters that are all the same sex. A large population might barely feel the effects of these extra deaths and missing births, but they could be catastrophic to a population the size of the ferret colony. Once the population has become all male or all female, for example, it is effectively extinct. If the flip of a coin represents life (heads) or death (tails), think of the probability of obtaining five tails in a row, or killing off an entire population of five, compared with the probability of obtaining 1000 tails in a row, or killing off an entire population of 1000.

The existence of a threshold population size helps explain why extinction becomes such a threat when contiguous habitats are fragmented into smaller patches. A patch has to be large enough to support a threshold number of individuals of any particular species, taking into account the

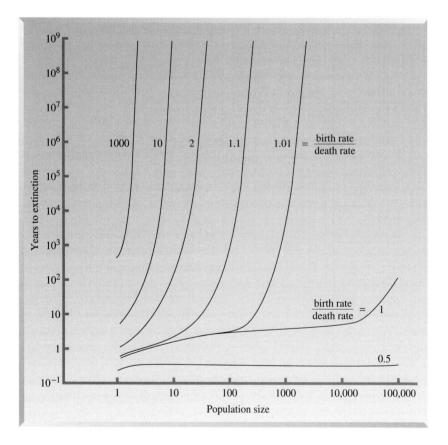

◀ How long a population can persist before becoming extinct depends both on its size and on the ratio of its birth to death rates. The numbers by the curves give ratios of birth rate to death rate for a birth rate of 2 young. Even a very small population can persist if its birth rate is much higher than its death rate; conversely, when the birth rate is half the death rate, even a large population doesn't stand a chance of surviving.

area requirements for individuals of that species. As a patch of habitat shrinks, the population of any species in that patch will decline in size, and its chance of extinction increase.

In general, when the birth rate is much higher than the death rate, a population will persist longer than it would if its birth rate were only a little higher than the death rate. That this should be so seems intuitively obvious. When births are numerous and deaths relatively fewer, a population soon reaches a density so high that it is impervious to the random clusters of bad luck predicted by demographic stochasticity. Suppose, however, that this ideal combination of high birth rate and low death rate is impossible to attain (perhaps a necessary resource is in short supply or fiercely competed for). One species could respond by putting most of its resources into breeding, so that many offspring are produced but adults die younger. Another species could put its resources into ensuring adult survival at the cost of producing fewer offspring. All other factors being equal, which is the best strategy? High mortality and high birth rate, or low mortality and low birth rate?

It may seem at first glance that the two alternatives should have an equal chance of success. But models incorporating demographic stochastic variability strongly favor the second alternative: for populations of similar size, with similar positive growth rates, species with low mortality and low fecundity will have a much higher probability of persistence.

Why should this be so? Put simply, if you live for a long time, you have more chances to gamble on reproducing at least once successfully. If you have a short life and can breed only once (albeit prolifically that one time), it is as if you are putting all your money into one rash gamble. The superiority of a low birth rate/low death rate over a high birth rate/high death rate is a consistent result of stochastic models and many computer simulations. Some supporting evidence comes from long-term studies of island bird populations. These studies have looked at the risk of population extinction, taking into account the number of nesting pairs and the survival rate of adults.

As was predicted, the studies found that the risk of extinction declined as the number of nesting pairs and the survival rate of adults both increased. To discern the effect we are interested in here, Stuart Pimm and Jared Diamond compared the extinction rates of large birds and small birds. Large birds such as gulls and birds of prey tend to have a higher survival rate than smaller species, and their birth rate is lower. Smaller species such as sparrows and wrens may have several broods in each year, but the stress of their higher reproductive activity tends to reduce adult survival. Thus the large birds represent the case of low birth rate/low mortality, while the small birds represent the case of high birth rate/high mortality. As the model predicted, populations of large birds show a tendency to go extinct less frequently than those of smaller species.

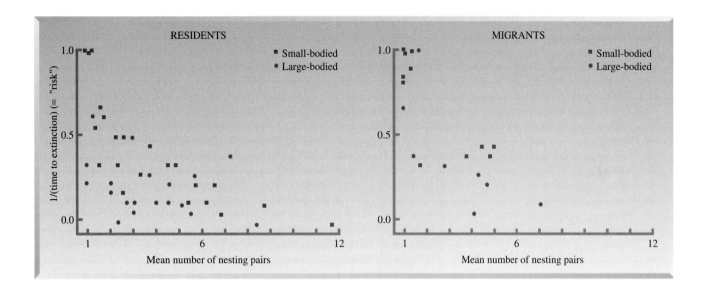

This effect makes an important contribution to the extinction debt phenomenon described earlier. Habitat fragmentation poses a disproportionate threat to larger species with larger area requirements, yet these are the same species whose high survival rates and lower birth rates give them an advantage compared with species having smaller body sizes and lower area requirements. Larger species will thus take longer to decline to extinction than we might otherwise expect, and the full impact of habitat destruction may not be realized until some time after irreparable damage is done.

The existence of a threshold population size, below which a population has a significant risk of declining to extinction, helped establish the concept of a minimum viable population size. Essentially, this concept assumes that for any species there is a minimum population size at which the species has a significant chance of persistence. (In fact, there is unlikely to be an exact threshold for any population, but rather an increasing probability of extinction as numbers increase.) The concept of minimum viable population size was first defined in 1982 by Mark Shaffer in a study of grizzly bears in the greater Yellowstone ecosystem. Shaffer formally defined a minimum viable population for any given species as the smallest isolated population having a 99 percent chance of remaining in existence for 1000 years. Unfortunately, the population sizes required to attain this ideal goal are usually in the thousands and unlikely to be characteristic of an endangered species, which in many cases will number in the hundreds.

To obtain accurate estimates of birth and survival rates, someone usually has to carry out a long-term study of a relatively undisturbed

▲ Stuart Plimm and Jared Diamond compared the risk of extinction (the mean reciprocal time to extinction in years) for 62 bird species on an island off the coast of Britain. Small-bodied birds seem to have a slightly greater risk of extinction than large-bodied birds, when the number of nesting pairs is the same. This is true for both year-round residents and migrant birds that visit Britain during the summer months.

▶ There are now fewer than 1000 grizzly bears surviving in the lower 48 states of the United States. Only the populations in Yellowstone National Park and Glacier National Park are likely to be viable through the next century.

population. Unfortunately, it may not be possible to study an endangered species in a location, or at a time, when its numbers are healthy. Instead, data tends to be collected at different times, or at a time when habitat alterations have already altered the survival and birth rates.

As an example of the false impressions that can be created by problematic data, consider the grizzly bears of Yellowstone National Park, which have been monitored since the pioneering studies of Frank and John Craighead in the 1950s. The total number of grizzly bears in Yellowstone declined in the late 1960s when officials closed the park's garbage dumps. A large proportion of the bear population consisted of bears who had grown up feeding at these dumps; unable to adjust to the dumps' disappearance, a large number of the bears died and the population declined. Since this abrupt decline, grizzly numbers have slowly increased as the backwoods bear population has expanded, so that there are now between 150 and 200 bears in the greater Yellowstone ecosystem. Biologists have collected most of their demographic data from bears that have been captured and fitted with radio collars. These data suggest that females have a litter of two to three cubs every three years, but that less than 20 percent of these cubs survive to become adults. Furthermore, a smaller proportion of males than females survive to attain sexual maturity. Any mathematical model we make of the Yellowstone grizzly population has to take into consideration these effects, as well as other details of how survival and birth rates vary with age and sex. These data on birth and death rates indicate that the population is declining. Yet strangely, when biologists make their yearly count of females seen with cubs, they find a slow increase in the numbers of grizzly bears in

Yellowstone. The difference between these two observations may reflect differences between bears with collars and those without collars. In particular, bears that have become habituated to humans are more likely to be collared than backwoods bears. These bears that approach camps and roads often suffer from higher mortality rates because they are struck by vehicles or shot when they get too close to humans.

These variations in data collection considerably complicate population viability analysis, since, as the grizzly example suggests, the best-quality data may be biased, whereas information available in lower quantity may be more accurate. Unfortunately, our ability to predict persistence for less well studied species is highly dependent upon the quantity and quality of information we have about the species. By definition, we usually have very little population data for rare or endangered species. This means that computer models of population viability should focus less on trying to predict when a population will go extinct, and more on trying to predict how changes in management might reduce the probability of extinction. In Yellowstone, for example, we could compare how the bears' population would hold up if we expanded the park by buying up areas of grazing land versus if we stopped excess logging on the west side of the park. To achieve such analyses, we have to consider environmental factors that cause variability in the survival and birth rates of a population.

Environmental Variation and Persistence Time

The calculations for population persistence become still more complex once we consider seasonal and yearly variations in the weather and food resources, and variations in the density of natural enemies such as predators and disease. In looking at demographic stochasticity, we assumed that, while a few individuals may be affected by chance events, the population's overall birth and death rates are constant. In most natural populations, however, these rates vary year by year and season by season. Changes in the weather, food resources, and so on lead to changes in mean survival and birth rates that affect all individuals in a population, not just a discrete few.

Consider the plant species *Astrocaryum mexicanum*. This palm, which grows in the tropical forests of Mexico and Central America, has a population growth rate only slightly greater than zero. Eric Menges of Archibold Biological Station in Florida has developed mathematical models for this species that predict its response to environmental variation. Menges's model assumed that seed production and adult survival would vary from year to year with swings in the weather. The model showed that the greater the variability in these two characteristics, the more dramatic the decline in a population's chance of persistence.

▲ *Astrocaryum mexicanum* is a spiny, feather-leafed palm that grows in tropical dry forests of Mexico and Guatemala.

The impact of environmental variation reveals itself clearly in its effect on the minimum viable population size. When Menges ran a version of his model that considered only demographic stochasticity, and simply ignored environmental variability, he found that a population of only 50 individuals could thrive with little risk of extinction. However, when he factored in the effects of a moderate amount of environmental variability, the population risked extinction unless it numbered at least 140 individuals. A high degree of environmental variability could be withstood only by a population of 380 individuals.

Why is the impact of environmental variability so much more striking than that of demographic stochasticity? The latter takes into account the fact that chance events will pick off the odd individual here and there. In contrast, a dramatic shift in the weather can place an entire community in jeopardy. The more likely such a shift, the more exposed the population to risk.

These results confirm some of the earlier theoretical findings of Dan Goodman and other workers, who showed for a variety of models that increased variation in population growth rate always leads to decreases in persistence. In a similar fashion to demographic stochasticity, persistence is more sensitive to environmental variation in mortality than to variation in fecundity. The simulations of Eric Menges suggest that plant populations are unable to escape the effects of environmental variation unless their annual population growth rates exceed 20 percent.

Unfortunately, very few ecological studies have run long enough to measure the real impact of environmental variation on the birth and death rates of even common species. Without such studies, predicting extinction times for endangered species remains a complex and incompletely resolved problem.

A possible solution to the problem of never having enough data to quantify the impact of environmental variation on birth and survival arises from the work of Joe Wright and Steve Hubbell. While using their models to explore the effects of subdividing a larger population into smaller, geographically separated subpopulations (a topic we return to in Chapter 7), they discovered an important additional property of populations: a statistic that described the level of variation in population size, called the "coefficient of variation in abundance," was a robust indicator of population persistence. Thus if we have information only on population numbers, we can estimate the coefficient of variation in abundance and obtain an estimate of persistence. As we have seen, large populations, which are more likely to persist, are not much influenced by demographic stochasticity; most of the year-to-year variation in large populations is a response to shifts in weather and other environmental factors. In contrast, small populations are influenced by both environmental and demographic stochasticity, and the effects of the latter become ever more dominant as population size decreases. Thus small populations, which are more prone to extinctions, are more variable in size from year to year and have larger coefficients of variation of abundance.

Curiously, Stuart Pimm and Ian Redfern, after reviewing long-term studies of animal and plant populations, have concluded that the coefficient of variation of abundance always increases with time. If this is an intrinsic feature of populations, it suggests that they inevitably contain the seeds of their own extinction. Alternatively, the seemingly universal rise through time of the coefficient may reflect the particular circumstances of this moment in life's history. All sets of data examined by Pimm and Redfern were collected in the late twentieth century, a time when pronounced changes to both atmosphere and climate, brought on by the activities of our own species, are influencing all living organisms.

Catastrophic Disasters

A tidal wave or large volcanic eruption can completely destroy the inhabitants of a tropical island. Indeed, large-scale environmental disasters may be vastly more drastic in their effects: there is some evidence to suggest that the impacts of large meteors falling on the earth may have produced clouds of debris that cooled the planet and led to the widespread extinctions observed in the fossil record. The unpredictable catastrophes that cause extinctions may be meterological (hurricanes, droughts), geological (landslides, earthquakes, volcanic eruptions), or political (wars, changes in land use). These events have a relatively low, but constant, probability of occurring at any time, and their impact on an entire community of organisms may be apocalyptic.

It has been difficult to obtain any empirical evidence telling us how frequently catastrophes occur. Nevertheless, their effects can still be accounted for in computer models of animal and plant populations: the researcher simply inserts random, large reductions in survival and birth rates into the models. The effects of these reductions are usually obvious: essentially, a catastrophe resets the population to a lower density whose size is determined by the magnitude of the crisis. The population then grows or declines from this density in a manner determined by its birth and death rates and the levels of stochasticity included in the model.

A classic example of devastation wrought by a natural catastrophe occurred when Hurricane Hugo managed to destroy in one night a large proportion of the remaining Puerto Rican parrot population and its nesting habitat. Luckily a few nests survived, but the recovery of this species from the disaster has been very slow. An important lesson was learned, however: unpredictable catastrophes present the strongest argument for saving several populations of a species (or community or habitat). If these

▲ This Puerto Rican parrot is from a species that came close to being eradicated by a single natural catastrophe, a powerful hurricane.

subpopulations are sufficiently isolated from one another, then a natural disaster that destroys one subpopulation is unlikely to hurt the others.

Social Behavior Disrupted

So far I have portrayed species as being at the mercy of outside forces, but a species' own behavior also has an important influence on whether a small population can persist. We have seen that many species are organized as metapopulations that exist on an array of habitat patches; each patch is repeatedly colonized and abandoned at a rate dependent upon its size and its distance from potential colonizers. Models have been developed to study populations characterized by this lifestyle. In particular, Russ Lande has developed a model for the northern spotted owl showing that a threshold number of habitat patches have to be occupied at any one time if the population is to persist. When the number of occupied patches drops below this number, or when the occupied patches become too dispersed, the population will decline to extinction. In some species patches of habitat may remain unoccupied because individuals only recognize suitable habitat when it is already occupied by a member of their own species. Ecologists studying seabird species have come up with a clever ruse to increase the number of island habitats occupied by these birds: they place artificial models of birds on the ground adjacent to suitable nest burrows, and real birds are soon settled on these formerly vacant sites.

The social organization of a species is another factor that plays a role in determining its chances of persisting. The individuals of some species (such as elephants and many primates) live together with groups of relatives; colonial species such as seabirds and fish aggregate in large groups containing both kin and nonkin. For these species, group living has decided benefits: a colony of seabirds can mob a predator until it gives up and departs; a large colony can search a greater area of ocean for the fish shoals on which the birds feed. When a bird returns to the colony bearing freshly caught fish, the others may capitalize on that bird's good luck by following the direction of its flight path back to the shoal.

The benefits of living in a group at first increase rapidly as the group size grows. Once the group gets above a certain size, however, the benefits begin to diminish. Perhaps there is no longer enough food in the immediate vicinity to feed all the individuals, and they must either eat less or go farther in their search for a meal. At this point tensions increase, fights break out, and the group may fission into two or more subgroups.

When the habitat of a group-living species starts to disappear or become fragmented, then the group in any patch of habitat may dwindle in number to the point where it has difficulty locating sources of food or defending itself against predators. There is thus a threshold below which the local population will decline rapidly to extinction. The tendency for a group that is too small to shrink away altogether is known as an Allee effect in honor of the ecologist W. C. Allee who proposed its existence in the early 1930s. Smaller colonies of seabirds, for example, are considerably less successful as breeders than larger colonies, in part because there are fewer parents to locate and mob predators, and in part because, with lower numbers, the birds can only search a smaller area of ocean for fish shoals.

Hopefully, the next generation of population models will be able to consider the details of the social life and behavior of endangered species. Simple stochastic models will always be useful in the majority of cases where we don't have much information about the ecology of a species. But because these models do not consider how animals change their behavior in response to shortages of resources such as nest sites, food, and mates, they only give us a coarse estimate of a species' likelihood of persistence over the next year, century, or millennium. If population biology has taught us anything in the last twenty years, it is that a species' behavior patterns are crucial in determining how a population fluctuates in response to the supply of natural enemies and potential competitors. Many classic examples of the mismanagement of natural resources stem from a failure to consider important biological relationships inherent in the population dynamics of animal and plant populations.

The diversification of life observed in the fossil record of the last 3500 million years did not take place through a random series of events. Whenever in the past new taxonomic groups have appeared and diversified into myriad new species, the appearance of these new species has coincided with the invasion of new habitats. These habitats offered new opportunities to organisms that could acquire the adaptations needed to exploit them. For example, the rapid evolution of flowering plants 100 million years ago created a crucial opportunity for the evolution of many new species of insect. As the human population comes to dominate the earth's terrestrial environment, it will increasingly become a potential resource for other opportunistic species. Our houses, barns, and workplaces have always provided a haven for rats, fleas, mice, and a whole host of bacterial species. As the human population has expanded, it has created wonderful opportunities for an increasing number of worm, protozoan, bacterial, and viral pathogens. These species are in the least danger of extinction, and many are expanding their populations as a direct consequence of the world's increasing human population. Recent scares about emergent pathogens, such as the Ebola and Hanta viruses, Lyme disease, and HIV, simply reflect nature's ability to capitalize on widely distributed and underutilized resources. As I shall discuss later, our ability to prevent our own exploitation by pathogenic microorganisms may lie with the plant and animal species of the tropical and temperate forests. If we continue to exterminate tropical species at the present unprecedented rates, this important potential will be lost.

When Is a Species Endangered?

Chapter 4

In 1973 the discovery of a small fish, in the Little Tennessee River, led to the suspension of the construction of a huge dam—the $116 million dollar Tellico dam in Tennessee. This was probably the first time in 10,000 years that a major project was halted for the benefit of another species. In this case, a new species of perch, the snail darter (*Percina tanasi*), had been discovered by a local biologist in stream waters that would have been flooded by the dam. As no other populations of the species were known to exist, the snail darter was legally classified "endangered"—that is, at serious risk of extinction—and construction of the dam was halted.

The legal battle over the fish was taken all the way to the Supreme Court, which ruled that halting construction of the dam was both legal and correct. The fight did not end there, however. Local politicians and business interests appealed to Congress, which in 1978 passed a bill that exempted the Tellico dam project from the rules protecting endangered species. Construction of the dam was completed, but luckily a number of other small popula-

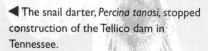

◀ The snail darter, *Percina tanasi*, stopped construction of the Tellico dam in Tennessee.

87

tions of the snail darter were found in streams that the waters behind the dam would not reach.

In ruling against a federal agency's plan to build a dam, the Supreme Court was acting according to the mandates of the U.S. Endangered Species Act, a law that prohibits the federal government from jeopardizing the existence of a species recognized as at risk of extinction. Opponents of the Act often use the snail darter as an example of how little known, and hence unvaluable, species can be used to halt economic development. Other critics of the Act assert that it gives priority to critters, rather than jobs, or that it has been woefully inadequate in protecting endangered species. In contrast, supporters of the U.S. Endangered Species Act regard it as an important example of the legal power that can be provided to species that are officially classified as endangered or threatened.

Despite the criticisms leveled against it, the U.S. Endangered Species Act has been a model emulated by the developers of both international legislation and corresponding laws in other countries. All countries have faced similar problems implementing their environmental legislation: lack of funds, opposition from landowners and business interests, tardiness in classifying species as endangered. In addition, the choice of species deemed deserving of protection often appears arbitrary: a distinct bias seems to be operating that favors species from certain groups, such as birds and mammals, over species from groups such as insects. If such laws are to continue protecting species from extinction, and ecosystems from destruction, they will themselves have to evolve; a crucial part of this evolution is the development of a scientific means of classifying species according to their risk of extinction. A new challenge confronts ecologists: how to present complex scientific arguments in such a way that they allow lawmakers to make informed decisions about which species need protection from human activities. Moreover, any arguments ecologists make will have to consider economic and political arguments in addition to the scientific ones. Ecologists must be able to compare the value of conserving a species with alternative human activities that might jeopardize conservation efforts, but, at least in the short term, appear to have political or financial benefits.

In this and the next chapter, we enter a curious new arena where the science of ecology and evolution meet the negotiation and compromise of politics, law, and economics. An increasing number of battles to protect biodiversity will be fought on legal and political grounds, rather than scientific ones. It is therefore crucial that we develop laws to protect species that are based on sound ecological and evolutionary principles. These laws should acknowledge that while other species provide services that make life possible for humans, the rights of these other species are not truly represented in a system where only one species votes.

Red Data Books

At present, there are two international organizations that define the status of plant and animal species throughout the world. The International Union for Conservation of Nature (IUCN) is made up of more than 600 government research institutions and private conservation groups concerned with the welfare of all species, while the Convention on International Trade in Endangered Species (CITES) is predominantly concerned with protecting species that are traded on the world market (except fish and whales, which are the subjects of special international legislation). The present system of classifying endangered species evolved from the use of the IUCN's "Red Data Books." When these books were established by the IUCN in the early 1960s, their role was to provide information on the demographic and geographical distribution of species and to focus attention on the plight of endangered species. As time passed, the Red Data Books began to set priorities and rank species by how much they needed the attention of conservationists.

Under the IUCN Red Data Book system, a species was classified as endangered, threatened, or vulnerable depending upon its perceived degree of risk. Although the task of classifying the world's numerous species often proved Herculean for the handful of scientists assembling the Red Data Books, they were very successful, at least for birds and mammals. Plants and invertebrates were less thoroughly covered, and this shortcoming provoked some biologists to suggest that, to ensure that these less glamorous species are also preserved, we should concentrate our efforts on saving particular endangered communities and ecosystems (although one way to do so is by saving species that characterize these communities and ecosystems, since one must usually save the habitat to save the species). Thus, there are laws in some countries that focus on the protection of habitats (e.g., Sites of Special Scientific Interest in Britain and Europe; the Wilderness Act in the United States), and there are laws in these and other countries that instead rely on lists of individual species to protect both populations and the habitats in which they live (e.g., the U.S. Endangered Species Act).

Many countries have relied on the Red Data Book format as they defined their own approaches to protecting endangered species. Whereas the format has served these countries well, its use occasionally produces anomalies. In particular, a broadly ranging and healthy species is sometimes perceived to be rare in a country that is situated at the edge of its geographical range. Avocets and ospreys are excellent examples. In the last fifty years both species have recolonized Britain, where appreciative naturalists have carefully guarded their nest sites and monitored their populations. In many ways, each species has inspired an increasing conservation

▶ The endangered northern spotted owl is one example of an "umbrella species." In preserving its habitat, environmentalists hope to save not only the bird but an entire ecosystem.

awareness in England (where the avocet is found) and Scotland (which the osprey prefers), yet neither species is endangered on a global scale.

Another problem was that the classification system used in the Red Data Books gave the most importance to highly endangered species, some of which were probably already beyond assistance at the time they were classified. Although conservation efforts may occasionally rescue a species in dire straits, more successful programs will focus on plant and animal species that we still have a reasonable chance of saving from extinction.

With its early emphasis on highly endangered species about which very little was known scientifically, the process of classification occasionally appeared arbitrary. Something better was needed to justify the sums of money that needed to be spent or the setting aside of habitat that could otherwise be logged or turned to profit in other ways. In many cases, weaknesses in the classification process made it easier for coalitions formed to promote exploitation of a species or habitat to develop arguments that became as influential in decision making as "expert scientific opinion" that takes into account the available ecological information about the species.

Environmentalists, lawyers, and politicians have called for more scientifically and politically acceptable ways of determining whether a species is endangered. Most agree, moreover, that laws to protect endangered species need to be implemented in such a way that they do not cause unnecessary financial hardship for people who own land that is habitat to a threatened or endangered species, or who run businesses that may threaten the species.

Conservation biologists are exploring several key questions that bear on how we classify species as endangered. Their first goal is to introduce more rigor into the classification process: Can quantitative guidelines be applied to classify levels of extinction risk for individual species? Their next goal is to analyze the effectiveness of the legal process: How quickly are species classified and does classification succeed in preventing extinctions? Finally, they must confront a probable, though unpleasant, reality: If, as seems likely, we are ultimately unable to protect all species, how do we determine which we should concentrate on saving?

Our legal attempts to protect endangered species have often been confounded by our ignorance, the fact that we know so little about most species, and even less about the roles they play in ensuring that ecosystems function efficiently. Because we lack the knowledge to pick and choose among species to save, we are presented with a dilemma: we can either create laws that try to save every possible species, or we can designate some species that require large areas of habitat as "umbrella species"—by legislating to protect these species, we will also protect the multitude of relatively unknown species that share the habitat we have characterized as essential for the well-being of our umbrella species. The northern spotted owl has been used in this way to protect the old growth forest of the northwestern United States.

The Northern Spotted Owl

The northern spotted owl (*Strix occidentalis caurina*) inhabits forests of Douglas fir in the northwestern United States. These forests, the rainforests of the temperate zone, are classified as "old growth" because many of the trees are several hundred years old, encrusted with lichens and other small epiphytic plants that help form a complex canopy layer. For decades bird watchers have known the spotted owl as an uncommon resident of the Douglas fir forests of Washington, Oregon, and northern California, but little else was known about the bird until the late 1960s. In 1967 an undergraduate at Oregon State University, Eric Forsman, taped the calls of these birds so that he could map their distribution in Oregon; his work developed into a Ph.D. thesis that revealed that spotted owls lived *only* in the old growth forests of the Pacific Northwest. Forsman showed that each pair of spotted owls requires a home range of 2600 hectares, of which around 1000 hectares must be old growth forest. Until 1990, the continued logging of these forests was rapidly reducing the number of habitat fragments that were sufficiently large to support the owls. That year, the Federal courts suspended logging until a plan for preserving the forest could be set in place.

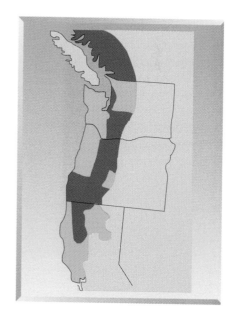

▲ The original range of the northern spotted owl stretched from British Columbia to the San Francisco Bay. The owl's habitat has been especially damaged, mainly by human activity but also by natural events such as fire and volcanic eruption, in the orange colored areas.

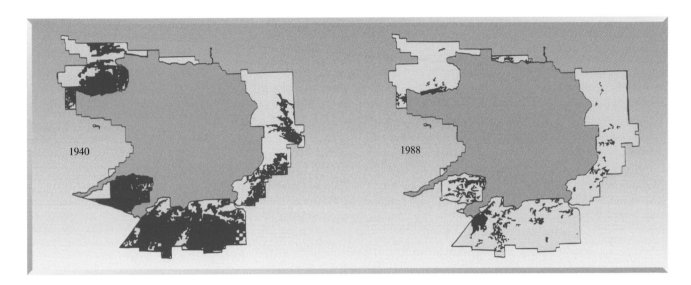

1940

1988

▲ The old growth forests of Washington State's Olympic National Park, the area at the center of the map, are still intact; there logging is prohibited. In contrast, the old growth forests (shown in dark green) of Olympic National Forest on the periphery of the park, where logging is permitted, have largely vanished in the years since 1940.

The owl's habitat needs were in direct conflict with the interests of the timber industry, which sought to replace the slow-growing, high-volume, and economically valuable old growth forest with young, rapidly growing timber stands that could be harvested every 60 to 80 years. Mostly since the 1950s, timber operations, fires, and the clearing of land for agriculture have reduced the old growth forests of the Pacific Northwest to less than 10 percent of their original area. As old growth forests contain a range of unique animal and plant species, including the marbled murrelet *(Brachyramphus marmoratus)*, the spotted owl has served as an "umbrella" species whose protection would ensure the continued existence of other, less charismatic, species.

The northern spotted owl is in fact one of three subspecies of spotted owl. There is also a California subspecies *(Strix o. occidentalis)*, which has less stringent dependency on old growth forests, and the Mexican spotted owl *(S. o. lucida)*. The timber industry has argued that the northern spotted owl is therefore not that unique, and that its protection could only be achieved at the cost of reducing timber production and a consequent loss of jobs in the local timber community. This argument ignores the fact that the majority of timber jobs will be lost once the forest is completely logged. At the rapid logging rates of the 1980s, most old growth forest would have disappeared by 1995, and the timber jobs along with it. A key feature of the Endangered Species Act is that it acknowledges that while people can be retrained, extinction is forever.

While attempts to save the spotted owl will cost timber jobs, they have created other jobs. However, most of these new jobs have been in different sectors of the community, particularly in the environmental and education fields, for people studying the birds and their ecology. Through these studies, biologists have developed more realistic models for species distributed as metapopulations. In addition, much of this research has illuminated more widespread debates on how to classify the risks of extinction for other endangered species.

The eventual listing of the owl as a threatened species has provided a classic test case of the U.S. Endangered Species Act. Lobby groups from either side have pitted the conservation of owls (and old growth forest) against the welfare of loggers and their families. The results from both long-term field studies of owl ecology and mathematical models of owl populations have been brought in to help determine how we might balance the owl's needs and the timber industry's. Calling on the predictions of population viability analyses, conservation biologists have developed plans that identify Habitat Conservation Areas (HCA), patches of habitat that, if they can be conserved, should suffice to maintain a viable population of the species even if other areas are logged.

Estimating the Risk of Extinction

We want a scheme for obtaining a dispassionate estimate of a species' risk of extinction, and that scheme should allow us to compare the absolute risks of extinction borne by species with very different ecological characteristics. We have already seen that population size, birth and mortality rates, and levels of environmental variability are all important in determining whether a population will perish or persist. When given sufficient information about a potential threatened or endangered species, we should be able to use population viability models like those introduced in the last

▼ Left: A species can be classified as safe, vulnerable, endangered, or critical by noting the probability of its declining by a specific percentage in the next fifty years. Right: Alternatively, a species can be classified by the probability of its becoming extinct within a certain number of years. The dashed line in the bottom graph corresponds to the fifty-year time interval assumed in the graph on the left.

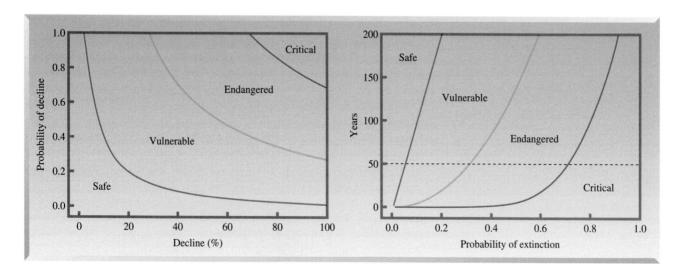

Mace and Lande's Criteria for Threatened Species

Population Trait	Critical	Endangered	Vulnerable
Observed decline	80% in 10 years or 3 generations	50% in 10 years or 3 generations	20% in 10 years or 3 generations
Geographical range	< 100 km^2 single location	< 5000 km^2 < 5 locations	$< 20,000$ km^2 < 10 locations
Total population	$N < 250$ $N_s < 50$	$N < 2,500$ $N_s < 250$	$N < 10,000$ $N_s < 1000$
Projected decline	$> 25\%$ in 3 years or 1 generation	$> 20\%$ in 5 years or 2 generations	$> 20\%$ in 10 years or 3 generations
Extinction probability	$> 50\%$ in 10 years or 3 generations	$> 20\%$ in 20 years or 5 generations	$> 10\%$ in 100 years

Note: N_s refers to the sizes of subpopulations that are found in different parts of the total range occupied by the species.

chapter to compute the likelihood of extinction. The problem is that the results may be difficult to compare from one species to the next; one solution is to define "isorisk" curves, like those illustrated in the graph on this page. The curves are drawn as lines that give a constant product of percent decline and probability; thus a population that has a 50 percent probability of declining by 10 percent in the next fifty years is assumed at similar risk to one that has a 5 percent chance of declining by 100 percent (to extinction!) over the same period. Because all the points on each curve represent a similar risk of extinction, by placing species on the appropriate curve we can categorize together species that are equally likely to go extinct, even though they have different population sizes and may be exposed to different threats. The curves can be used to demarcate regions on the graph that define populations quantitatively as safe, threatened (or vulnerable), endangered, or critical.

Is it possible to classify species in this way? That is, are we able to calculate a species' probability of decline over the next 5, 10, or 50 years? Obviously if we have lots of information about the species the calculation will be possible, but endangered species are often rare and insufficiently studied. Instead, we usually have bits of anecdotal information collected from studies of the species behavior or physiology; these studies may have made no attempt to understand the demography of the species. In many cases, we will only know the barest details about the social system, geographic range, and feeding habits of the species. Is it possible to use any of this information to say anything about population viability?

Georgina Mace and Russ Lande have developed a more rigorous way to classify threatened species using the sort of information that is likely to be available. The categories that result roughly correspond to the categories of threat illustrated in the graph on page 93. Mace and Lande's approach has been modified after widespread discussion in the scientific conservation community so that it takes into account the type of data that researchers actually collect from studies in the wild. The criteria now form the basis for the new IUCN categorization, called the Red List. These new criteria explicitly acknowledge that because different researchers collect information of different types (in order to address different questions) it is rarely possible to obtain similar information for different species.

Mace and Lande's criteria represent a brave attempt to weigh the available information in light of the different factors determining the risk of extinction, ranked in order of importance. The criteria are based on a number of indices of population size, and change in population size, all of which should be possible to measure in three to five years' detailed field work. The ideal research project would measure five specified criteria, although meeting any one of the five criteria qualifies a species for listing at that level of threat. The five criteria used to classify a species as at risk are (1) a substantial rate of decline, which could be observed, inferred, or projected; (2) a small population that is either single or fragmented into several isolated subpopulations; (3) a small geographical range that seems to be shrinking, or is projected to shrink in the future; (4) a strong probability of extinction within a specified time period, revealed by a quantitative population viability analysis. The criteria are listed in the table on the facing page; at first sight they appear fairly complicated, but notice that the first row corresponds to a rate of decline that is directly analogous to the "isorisks" described above. There are then a number of population measures that serve to place a species in one of these classes of risks.

The best way to grasp the usefulness of these criteria is to apply them to our closest evolutionary relatives: the world's primate species. There are over 200 species of primate, ranging from the tiny mouse lemur (*Microcebus murinus*) of Madagascar to the great apes, which include the mountain gorilla (*Gorilla gorilla*) and our closest living relatives, the chimpanzee (*Pan troglodytes*) and the pygmy chimpanzee (*Pan paniscus*). Roughly 90 percent of all primate species live in the tropical forests of Asia, Africa, and Latin America; in all these regions their populations are decreasing. The majority of these species have populations that are smaller than most human cities and towns, or even many colleges and universities.

According to the criteria described above, used to create the IUCN Red List, one out of every three of the world's 200 primate species is threatened with extinction, and one in seven is so endangered that it could be extinct early in the next century. The disappearing primates are, in the main, victims of the destruction of their tropical forest habitat, although

▼ Like most primate species, the total world population of chimpanzees is smaller than that of a small city. Many primate species have total populations smaller than many universities or colleges.

primates are also hunted, mainly as a source of food, but also as a source of subjects for medical research.

How many of the world's primate species may be classified as endangered, threatened, or vulnerable? An initial attempt to address this question suggests that approximately 12 percent of the world's primates are at critical risk of extinction, a further 12 percent are endangered, and 14 percent are vulnerable. More than 10 percent of these species are concentrated in Madagascar, while the rest are divided almost equally among Africa, Asia, and Latin America. Unfortunately, merely categorizing these species as endangered has done little yet to reduce their decline. Yet it does allow us to identify areas of habitat that are especially worth saving because of the numbers of endangered primate species they harbor. To successfully achieve this goal, our biological classification of threat must somehow be transformed into an environmental policy that provides legal protection for these species.

The U.S. Endangered Species Act

The U.S. Endangered Species Act, signed into law by President Richard Nixon in 1973, has been described as being "to endangered species what Social Security is to old people" (National Public Radio environmental correspondent John Nielson). The decision to list a species as endangered is made by the Secretary of the Interior, following the recommendation of the Fish and Wildlife Service, when any one of five conditions is met: (1) the destruction of its habitat is in progress or threatened; (2) the species is being overexploited (individuals are being killed faster then they can be replaced) for commercial, recreational, scientific, or educational purposes; (3) it is suffering losses from disease or predation; (4) existing laws and regulations are inadequate to protect the species; (5) there are "other natural or manmade factors affecting its continued existence." Once listed, a "species," which for the purposes of the Act is assumed to include subspecies and populations, may not be harmed, nor may federal agencies authorize, fund, or carry out any action that is likely to jeopardize its continued existence. Here "harm" has been broadly defined to include the destruction of habitat deemed necessary for the viability of the species.

At present a two-tier system, modeled on the IUCN Red Data Books, classifies all listed species as either "endangered" or "threatened." An endangered species is "any species which is in danger of extinction throughout all or part of its range"; a threatened species is "any species which is likely to become an endangered species within the foreseeable future."

The species currently listed as threatened or endangered, both internationally and in the United States, are unevenly distributed among the different taxa. As the figure on the facing page reveals, there seems to be a bias toward mammals and birds; our greatest failings to list are among the invertebrates and, on an international scale, among the plants. In some ways, the listed species reflect human fascination with the mammalian species that are most closely related to ourselves, as well as the relative numbers of biologists trained to work on different taxa. On the other hand, the imbalance probably also came about because officials deliberately listed species with large habitat requirements so that they could act as "umbrella" species to help protect specific ecosystems.

Once a species is listed as threatened or endangered, the U.S. Endangered Species Act mandates that a recovery plan be developed to restore

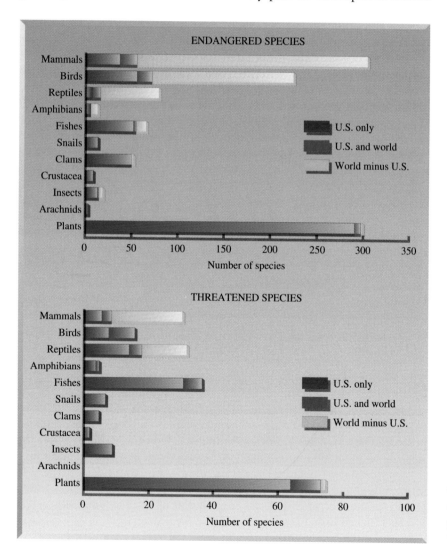

These graphs show the total numbers of species listed as endangered or threatened in the United States and the rest of the world, as of April 2, 1993. The species are divided into the administrative taxonomic groups used by CITES and the U.S. Fish and Wildlife Service.

each species to a level where it is believed safe from extinction. The recovery plan identifies steps that could be taken to prevent further habitat loss. The plan for the loggerhead turtle, for example, requires that almost 90 miles of Atlantic beachfront property where the turtle nests be put in public ownership, whereas that for the blunt-nosed lizard calls for all conflicting land uses to be reduced or eliminated on 60,000 acres of the San Joaquin Valley in California. In addition, the plan sets population sizes that have to be achieved before the species can qualify for downlisting. A relatively high proportion of listed plant species have recovery plans, while only a small proportion of listed mammals and birds have such plans. The enormity of the task is such that the small number of Fish and Wildlife Service officials faced with constructing the recovery plans cannot keep up.

The taxonomic distribution of species listed strongly reflects human perceptions of the "value" of different species. It is thus biased toward the aesthetic vertebrates and butterflies, and away from the invertebrates that form the greater fraction of biodiversity in a region. Stephen Kellert of Yale University threw some interesting light on this problem when he surveyed a large sample of randomly selected citizens and asked them which of a list of endangered species they would favor protecting, even if the result was higher costs for an energy development project. Most people were prepared to make some personal sacrifices for species such as the bald eagle, mountain lion, Agassiz trout, American crocodile, and silverspot butterfly; but they were not prepared to make sacrifices for the eastern indigo snake, cave spider, or a plant such as the furbish lousewort. Yet these little regarded plants and invertebrates are crucial for ecosystem processes such as nutrient cycling, pollination, pest control, and seed dispersal—and certainly most species of plants are relied upon as sources of food by some organism.

Problems with the Listing Process

The success of the Endangered Species Act has been disappointing to some. At present, its efficacy is confounded by the taxonomic biases described above and by two additional problems: the slow rate at which species are listed, and the size of populations when listed. These last two problems are connected, since the longer it takes to list a species, the smaller its population size is likely to be when finally accorded legal protection. Since the Endangered Species Act was signed into law in 1973, an average of about 40 species have been placed on the endangered list each year. The number of listings far exceeds the number of delistings: while 955 species had been placed on the list by the end of 1994, only 18 species have been successfully moved from endangered to threatened, or have been

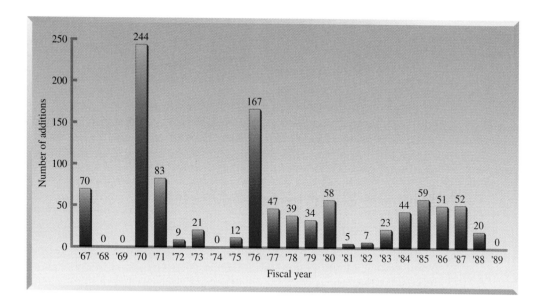

▲ The number of species listed under the U.S. Endangered Species Act per fiscal year, as of May 1, 1988.

moved off the list entirely, and more than 3000 species are candidates for listing. Unfortunately, some species were removed from the list when they became extinct. The Tecopa pupfish, the first species on the list to be lost, was found only in the South Tecopa Hot Spring in Inyou County, California. This species was remarkable for thriving in the highest water temperature ever recorded for a fish (up to 40°C, or 104°F). Its sole habitat was destroyed when a bathhouse built on the hot spring contaminated the water. Simple modifications to the bathhouse could have saved the species.

Unfortunately, the size of many animal and plant populations when finally listed does not bode well for their chances of recovery. An animal species tends to be listed when its average total population size is around 1000 individuals (divided among two or three subpopulations), while plant species are listed at an average population size of just over 100 individuals (usually divided among four subpopulations). When threatened and endangered species were compared, endangered vertebrates had been listed at population sizes of around 400 individuals, while threatened populations had been listed at population sizes of around 4000 individuals. Similarly, endangered plants had populations of fewer individuals than threatened species (99 versus 2500), and fewer populations (3 versus 9) as well. In all cases, population sizes at the time of listing were below those recommended by the IUCN Red List as qualifying for international endangered status (see the table on page 94).

The detractors of the endangered species act point out that only five species have recovered to the point where listing is no longer necessary. This is a misconception; in fact, 18 species have been downgraded to a lower class of risk, and many species are still in existence that might otherwise be extinct. Nevertheless, this success rate is still modest and may reflect the small sizes of populations when finally listed as endangered or threatened. The legal protection afforded to endangered species may not have any real effect until species are protected at larger population sizes. As it costs more to try to save a small population, while the chance of success is diminished, it makes sense, on both economic and demographic grounds, to get species placed on the endangered list sooner.

Spending on Endangered Species

A second criticism of the endangered species act is that it costs significant amounts of money to protect each species. One congressman has been heard talking of the billions of dollars it costs to save each endangered species. As recently as 1991 the total amount appropriated to the Fish and Wildlife Service to administer the act was less $38.7 million; the net total since the act was signed in 1972 is less than $337 million (the cost of a single jet fighter). The amount of money allocated by the states for "reasonably identifiable expenditures made primarily for the conservation of endangered and threatened species" for fiscal year 1991/92 was $131 million. The total of $170 million, in both state and federal funding, is plainly less than a billion dollars for each species, although more than half of this money was directed toward the preservation of just one species in one state—the red-cockaded woodpecker in Florida. The $64 million spent on this species represents the one-time cost of buying land to create reserves for the woodpecker. The mean amount spent on each of the remaining listed species was around $104,000. This figure is a factor of 5 higher than the commonest amount spent on a species of around $20,000 per year. All these figures are considerably less than the billion dollars spent every two days by the Pentagon; indeed the total budget is only half the cost of each jet fighter among the three or four the U.S. military crashes each year on "training" missions.

The Office of the Inspector General has estimated that to achieve the recovery of all species listed in 1990 would cost around $4.6 billion, over a period of ten years. The Fish and Wildlife Service budget for recovering endangered species in 1995 was $70.4 million. The contributions of other agencies toward protecting endangered species raise the total to just over $100 million.

Many other national conservation organizations, such as the Kenya Wildlife Service, are also inadequately funded to develop recovery and monitoring plans for the large number of species that need them. Even moderate funding in this area could create jobs in wildlife management that require skills that are needed at both the national and international level.

Unfortunately, the Fish and Wildlife Service does not have a fixed set of rules for deciding how much to spend on a species. The meager budget and the complications of assigning priorities to species mean that the money spent in any one year is very unevenly distributed. More than half of the money alloted for 1990 went to 11 species, while 114 listed species received no funding.

Ironically, only a very small proportion of the money spent on endangered species goes toward research. Most of the state and federal funds are used for the purchase of land and to defray legal costs. The annual amount of money spent on environmental legislation of all types in the United States is around $9 billion, and less than $1 billion of that is spent on all environmental research. In some cases the scientific information we require to make a rational decision on how to manage a species is unavailable because scientists have been denied permission to work on an endangered species, lest their activities put the species into further jeopardy!

Private landowners are the largest group of people who occasionally suffer from the administration of endangered species legislation. In the United States, the presence of an endangered species of bird, the California gnatcatcher, has thwarted building developments in coastal areas and consequently reduced the value of several privately owned patches of land. In East Africa and India, elephants raid crops and destroy buildings, but are legally protected from landowners who might seek retribution. In both these cases, landowners could be compensated; in the case of the elephant, such compensation could come directly from revenue generated by tourists who come to see the animal. Furthermore, in many places the provision of tax deductions and reduced inheritance taxes could actively encourage landowners to preserve habitat for wildlife.

The cost of actively protecting a species fails to consider the loss of income that would have been generated had the habitat been put to other economic uses. To assess the extent of this loss, we need to know how many federal activities (including private development activities that require a federal permit) have been stopped as a result of "jeopardy" rulings under Section 7 of the Act. Jeopardy rulings prohibit federal agencies from authorizing, funding, or carrying out any action that is likely to jeopardize the continued existence of a listed species. Out of the nearly 50,000 activities evaluated during the period from 1976 to 1986, more than 99 percent were found not to jeopardize the species in question, or were capable of slight modification that minimized their jeopardy.

Where jeopardy rulings are brought against an activity, the agency can appeal to the so-called "God committee" for an exemption. The "God committee" was formed as a result of the Tellico dam and snail darter controversy, which resulted in Section 7 (the jeopardy clause) being added to the Act. The "God committee" met in 1992 for the first time in 13 years to consider an exemption for timber harvests that were in conflict with the habitat needs of the spotted owl. The federal government is now implementing a plan to preserve the owl that allows logging on only one-fifth the scale seen in the 1980s, although opposition from Congress may yet put the plan in jeopardy.

The complexities of funding become even more byzantine when we move from a consideration of private to public lands—areas that are not specifically owned by one person but can be used by many. In the United States these areas come under the jurisdiction of the Bureau of Land Management (BLM) and, where forested, under the U.S. Forest Service (USFS). These federal agencies manage more than 20 percent of the landmass of the United States; the huge areas of natural habitat within their domain provide the last refuges for many endangered species. At least 45 percent of federally listed endangered species in the United States have some, or all, of their range on federal lands, and at least 8 percent are found exclusively there. Unfortunately, the laws developed in the nineteenth century to encourage cattle ranching, mining, and logging are still applied today, and these activities usually lead to a net financial loss to the federal government. For example, a recent study by Elizabeth Losos and colleagues at the Wilderness Society showed that in 1992, of the 120 national forests 95 operated at a loss totaling $174.9 million, because the forest service spent more money preparing and administering timber sales than it received in revenue. Similarly, the BLM spent $49.8 million in 1990 to operate its grazing program, but collected only $19.3 million in fees. The full irony of this situation becomes apparent when we realize that the federal government has to spend additional monies to protect the endangered species put at risk by the subsidized activities. In many parts of the United States the federal government is effectively paying one group of people to damage public lands, while paying another group of people to repair this damage. Characteristically, the group that receives the largest and most direct subsidy (the ranchers, foresters, and mining industry) complains most vociferously that the activities of the conservation community constrain its economic well being!

In a recent survey of species officially listed as endangered in the United States, Losos and colleagues showed that 479 (66 percent) of the 777 species on the list were threatened, at least in part, by grazing, hardrock mining, logging, water development, recreation, or some combination of these. Water development (dams, flood control efforts, etc.) threatened the greatest number of species (29 to 33 percent) and recreation the second greatest number (23 to 26 percent), while grazing (19 to 22 percent) and timber extraction (14 to 17 percent) also had a large impact. Although mineral extraction only harmed 4 to 6 percent of endangered species, at least $1.2 billion worth of minerals were extracted from federal lands in 1992 with no compensation paid to the government. If the federal government had charged the 12.5 percent fee it levies for gas and oil taken from federal lands, it could have easily recovered the cost of the endangered species program for the entire country in that year.

Which Species Should Receive Priority for Listing?

A total number of 1479 species worldwide were listed as endangered by the U.S. Fish and Wildlife Service on January 1, 1995—approximately 0.1 to 0.4 percent of the world's known species (if we assume there are only 1.5 million species). If the estimates are correct in suggesting that we are losing as many as 1 percent of the world's species each year, then plainly we have failed to protect a large number of the species at risk. If the complexities of listing species as endangered mean that only some of the potentially endangered species can be legally protected, is it possible to determine which species we should concentrate on to ensure that the most biodiversity is conserved?

There is already one important and much used way of selecting which species to protect: list those with large habitat requirements. Species such as the spotted owl, the grizzly bear, and the African elephant need many square miles of forest or savanna, so by protecting viable populations of these species we can protect whole ecosystems containing a wide diversity of less-charismatic species. Unfortunately, conservation programs that concentrate on individual species are frequently subject to superficial economic calculations by opponents of conservation that put a dollar value on each surviving individual in the population and juxtapose this cost against the number of jobs lost in an extractive industry such as forestry or ranching. It is obviously easier for people in welfare industries such as mining, forestry, and ranching to rely on this form of argument than to think about the long-term costs of their actions.

An alternative approach is to ask whether some species are more "deserving" of protection—do they have some quality that makes them especially worth preserving? Of course, every species is by definition unique in some way, and the possible criteria for being deemed worthy of protection are vast in number, but one proposed approach is to examine the relationships revealed by the evolutionary tree for the group of organisms we are interested in conserving. These relationships can be used to support different criteria for listing species: some scientists suggest placing greater emphasis upon taxonomic rarity when classifying endangered species, as we do with the giant panda, which is unique among the bears. Others suggest that we should concentrate on saving groups of closely related species that are evolving rapidly, such as the honeycreeper finches of Hawaii, a closely related group of birds that have evolved in the last six million years into 42 species that feed on flower nectar. Trying to make these judgments objectively is of course totally irrational; it is asking us to compare the few works of Leonardo da Vinci with the many works of Picasso, and then telling us we are only allowed to keep the works of one artist, because the other artist's work will be burnt to keep a small army warm at night.

Taxonomy as Destiny

As an example of the approach that values taxonomic rarity, consider the tuatara, an iguanalike reptile that lives on islands off the coast of New Zealand. The species is the sole survivor of a once widespread group that lived more than 200 million years ago in the Triassic period. The tuatara is essentially unrelated to present-day lizards, despite a superficial resemblance. Its most distinctive feature is the presence of a partially developed eye in the back of its head; most other vertebrates have retained this organ in vestigial form as their pineal gland.

Three species of tuatara were described in the nineteenth century; one has since gone extinct, and of the two that remain, *Sphenodon punctatum* and *S. guntheri*, only the commoner *S. punctatum* is entitled to protection under the 1895 Animals Protection Act in New Zealand, because only *S. punctatum* was recognized as a species when the law was passed. Since recent research has provided ample evidence to show that *S. guntheri* is a dis-

▶ The species *Sphenodon punctatum* is one of two species of tuatara, the sole relicts of an ancient group of reptiles.

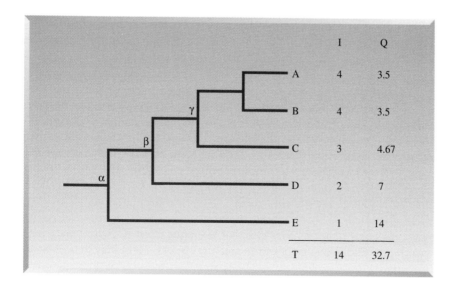

	I	Q
A	4	3.5
B	4	3.5
C	3	4.67
D	2	7
E	1	14
T	14	32.7

◀ A hierarchical classification, or evolutionary tree, for five imaginary species labeled A, B, C, D, and E. Different approaches can be adopted for weighting the species by their evolutionary uniqueness. One approach gives equal weight to each sister group, defined as having branched from a common node. An alternative approach considers the amount of information ("I") present in the tree. The relative taxonomic uniqueness of a species ("Q") is calculated by dividing the total information score by the species' score.

tinct species, should both species now receive protection? Or, to take this argument further, do these two ancient species of reptile deserve more protection than more recently evolved species of reptile? To continue our earlier analogy, we can consider the tuatara as the herpetological equivalent of the Lascaux cave paintings, while all other lizards represent the entire canon of European art since the Renaissance.

To address this question we need information about the evolutionary relationships of the different reptile species. The figure on this page illustrates a hypothetical tree for five species. The tautara would be roughly equivalent to species E, while all other major lizard groups would be represented by the other branches of the tree. One approach to measuring taxonomic distinctness is to give equal weight to each sister group (those groups that could be defined to have branched from a common node). In our hypothetical tree the node alpha defines species A, B, C, and D as one sister group, while species E forms another. Similarly, node beta defines species A, B, and C as a sister group, and node gamma defines species A and B as a sister group. By this approach, species E is weighted equally to the total value of A, B, C and D; species D is weighted equally to the total value of A, B, and C; species C is weighted equally to the sum of A and B. If we applied this approach to the tautara species, they would be weighted equally with all 6800 other reptiles. The opposite approach would be to weight each species as unique, in which case the tuataras would be equivalent to any individual species of skink, lizard, or snake.

Plainly, a sensible intermediate between these two extremes is needed. One approach is to consider the amount of "information" present in the evolutionary tree in weighting the species. For example, the cladogram (or evolutionary tree) lets us make four taxonomic statements about A (it belongs to groups AB, ABC, ABCD, and ABCDE); in contrast, D belongs to only two groups (ABCD and ABCDE). A total of 14 (4 + 4 + 3 + 2 + 1) information statements may be made about the tree. The relative taxonomic uniqueness of each species is then calculated by dividing the total score by its individual score. Thus the taxonomic uniqueness of A is 3.5 (14/4), of C is 4.67 (14/3), and of E is 14 (14/1). This method still gives extra weight to species in ancient lineages, but in a less extreme fashion than using sister groups for weighting. Although the approach requires a fairly well resolved cladogram to work, it provides a useful method of calculating an index of taxonomic rareness. When applied to the tuataras it suggests that both species are equally worthy of conservation attention. However, this technique is likely only to be useful in cases where we need to demonstrate the value of saving a species that is unique from an evolutionary perspective; it takes no account of the role the species might play in ecosystem function.

Nevertheless, an important virtue of this technique is that, besides pointing to species from ancient lineages that are especially worthy of protection, it can be used more powerfully to highlight whole geographic regions where conservation action is needed. The key is to consider the relative taxonomic uniqueness of the species in different regions. This technique has been applied, for example, to create a map ranking the areas of importance for conserving bumblebee species. The map on the facing page shows that the first four regions of importance are located on different continents. Although the area with the greatest number of species is in Ecuador, the 10 species found in this region contain some species that are closely related, so they contribute less than 15 percent of the total weighted diversity score. In contrast, the 9 species from the Gansu region of China have diversified into less closely related species that contribute almost 23 percent of the total bumblebee diversity. The first geographic region to complement the Gansu region, in the sense of having the bumblebee collection that is least similar, is the Big Horn region of North America. The Big Horn region has only 4 species, but adds 15 percent to the total taxonomic diversity represented. The third region to complement Gansu is then Ecuador, which shares no species in common with Gansu or the Big Horn. These three areas capture more than 50 percent of the world's total bumblebee diversity. If this technique were to be applied to species from a range of taxonomic groups, it would be possible to build up maps of areas where many unique species live.

Two criticisms of these approaches are that they assume that all the branches in the tree are of equal length (essentially they use trees that ignore the time since species groups split from each other) and that they assume that maximizing diversity is the goal of all conservation efforts. The first criticism may be met in theory by modifying the algorithms to con-

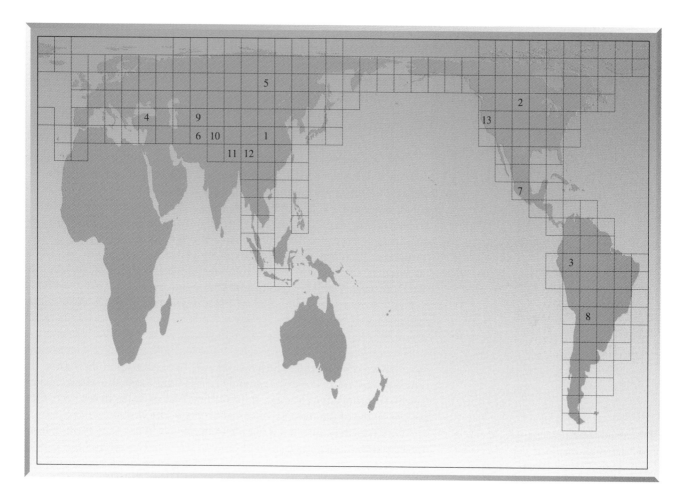

▲ This map ranks areas of the world by their contribution to bumblebee diversity. The Gansu region of China, marked number one, contributes almost 23 percent of the world's bumblebee diversity.

sider the length of the branches for each tree. In practice, however, modifying the algorithms hasn't been possible since it requires fully resolved evolutionary trees for each group we are trying to conserve. As yet, evolutionary biologists have not even reached consensus on a tree for the mammals!

The second criticism has been made by Terry Erwin, who argues that the major goal of conservation is to allow evolution to continue its course. In Erwin's view, it is best to concentrate on saving the diverging twigs at the tips of the branches, where rapid evolution may still be occurring. Ancient, relict species that have remained unchanged for millions of years will be of limited value under this scenario because they have ceased to evolve. This position may at first seem extreme: it essentially says we don't need all those Leonardo da Vinci's, Bellini's, and Matisse's because there are plenty of young artists around! Nevertheless, there are several important examples of taxa where it may be important. For example, recent molecular genetic information has shown the cichlid fish of the African great lakes have diversified into new species very rapidly over the last million years or less. The 40 species in Lake Victoria all evolved from a single common ancestor, while the hundreds of species in Lakes Tanganyika and Malawi have probably evolved from two or three ancestral species. All of these fish species are threatened by the introduction of the predatory Nile perch into these lakes and by overfishing.

The strengths and weaknesses of these two approaches can be made clear by briefly considering the taxonomic distribution of the world's primates. The taxonomic uniqueness approach would have us concentrate on saving representative examples from groups that diverged from the evolutionary tree at an early stage and contain only a few species. Thus pongids (chimpanzees and gorillas), prosimians, and lemurs would receive disproportionate attention; less effort would be expended upon each species of the large cercopithid, cebid, and colobine families. In contrast, the alternative approach—which concentrates on comparatively rapidly evolving groups—would boost attempts to protect the cercopithids, cebids, and colobines and reduce efforts for the pongids, prosimians, and lemurs. Both approaches have the advantage that they may focus valuable conservation dollars and produce a few successes; however, both approaches ignore the fact that we can't predict which species will be successful in the future. To return to our art analogy, we can only be thankful to people who kept Van Gogh's paintings, for although the paintings had no value in his time, they are now some of the most expensive artworks in the world.

Unfortunately, only a few countries in the world have any form of legal protection for endangered species. Only in western Europe and the United States is the establishment of rights for nonhuman species following the same path from heresy to policy that equal rights for women and equal racial rights followed earlier in the century. Furthermore, throughout the world the present system of listing requires biologists to assume the burden of proof to demonstrate that a species is in trouble. Thus, many species that have not been seen in the wild for more than fifty years are still assumed to be in existence, because they have not been proven extinct. Plainly, we have a long way to go before the full force of the law can be used to help conserve biodiversity.

▼ Four species of cichlids are found only in Lake Tanganyika (on the left) or Lake Malawi (on the right). The spectacular evolution of this family produced hundreds of species within single lakes. The two *Tropheus* species shown on the left live only over rocky habitats. Long beaches or estuaries forming barriers between these habitats probably kept the ancestral *Tropheus* populations separated and allowed them to evolve into new species.

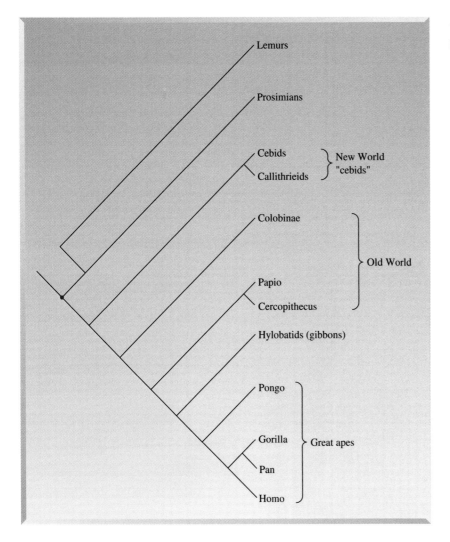

◄ An evolutionary tree of the major primate groups.

Finally, let us return to the old growth forests of the Pacific Northwest. The level of logging sustained in the 1980s would soon destroy all the region's forests, taking the timber jobs with them. Why didn't the timber industry remove a consistent and limited yield of timber that would allow the forest to continue? Indeed, why aren't foresters throughout the world harvesting trees at a slower, more sustainable rate? Forests, of course, are not alone in being exploited for trade; fish, birds, and other animals are removed from the wild to become commodities on the market. The next chapter will examine why so many of these species quickly turn up on the lists of endangered species.

Wildlife in the
Marketplace

In October 1990 a group of Norwegian scientists dissecting the carcass of a pregnant blue whale they had captured were amazed to find that the unborn calf was a hybrid between a blue whale and a right whale. Their discovery—the first recorded example of hybridization between two whale species—was greeted by the team as an important justification for "scientific" whaling, the capturing and killing of whales in order to monitor their populations. Unfortunately, the hybrid was probably a sign of the continued imperiled state of the great whale species. For well into this century whales have been exploited by humans; by the 1980s, blue whale populations were at such low levels that matings with males of a different species may have been the only opportunity presented to fertile females to reproduce.

Whales were at the center of a major industry in the opening years of the twentieth century. The whaling industry employed large numbers of people and supplied resources that had a wide variety of uses from food stuff to lubricant to fuel oil for lamps. The

◀ A fur merchant displays two snow leopard pelts in a market in Kashgar, China. Economic factors make the over-hunting of species like the snow leopard almost inevitable.

▶ Sperm whale oil, stored in casks on the way to the factory, was used in the manufacture of a number of nineteenth-century products, from cosmetics to soap to candles to margarine. It was even an ingredient of the explosive nitroglycerine. Even in our own day, sperm whale oil has been used to make germicides and detergents.

populations of many whale species were decimated in the industry's zeal to increase its harvests. Today, whaling is undertaken by only a few nations and then mainly by native peoples or for "scientific" purposes— although the killing of whales in the name of "science" is in fact a thinly disguised means of supplying whale flesh to the lucrative Japanese sushi market. Although a few species of whale have been able to recover from the brink of extinction, the populations of most species remain seriously depleted. Of the more than 200,000 blue whales that originally populated the southern hemisphere, for example, there are now estimated to remain as few as 1200.

At present there are vociferous calls for more widespread "sustainable" use of whales and other biological resources. According to this point of view, these resources should be used in a such a way that there is no net reduction of future generations. On the other hand, the so-called "wise use" movement believes that natural resources should always be utilized for profit without regard to their future. Still others argue that natural resources have an intrinsic value that is not readily quantified on an accountant's spreadsheets. It turns out to be extremely difficult to achieve sustainable use of species taken from the wild. Economic factors encourage overexploitation, and government regulations intended to prevent overexploitation may inadvertently promote it. The story of the rise and decline of the whaling industry illustrates the typical problems that arise, for reasons both economic and ecological, when natural populations are exploited for profit.

The Sorry History of Whaling

The discovery that sperm whales provided a rich source of lubricant oil, in addition to a source of candlewax and lamp fuel, ushered in the golden age of Yankee whaling at the beginning of the eighteenth century. By 1715 six small sloops were engaged in deep sea whaling out of the port of Nantucket, Massachusetts. Although the Portuguese and Basques had begun whaling off the coast of Newfoundland in 1535, the industry now expanded rapidly. To satisfy the demand for sperm whale oil, whalers hunted population after population of sperm whale to collapse. They also hunted species such as the right, humpback, and gray whales as the whaling fishery steadily expanded throughout the eighteenth and nineteenth centuries.

In those earlier days of whaling, men in small boats, launched from a sail-powered mother ship, had to row close enough to the whale to drive in their harpoons by hand. Modern whaling began in the 1860s, when the Norwegians launched their first steam-powered whaler equipped with a newly invented harpoon gun. At that time the stocks of right, humpback, and gray whales off the coasts of Europe, Labrador, and New England were beginning to decline, and with the invention of steam-powered vessels and harpoon guns whalers could pursue the swifter blue, fin, and humpback whales that preferred deeper water. By the beginning of the 1900s, stocks of whales had become depleted throughout the Atlantic, as

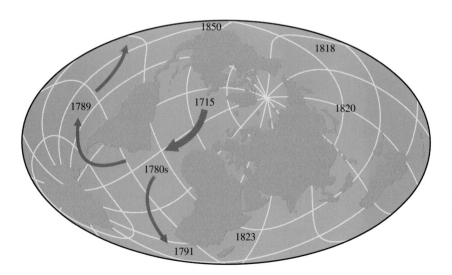

◀ Sperm whalers traveled ever farther around the globe in their search for new whaling grounds. The map shows the years in which the major sperm whaling grounds were discovered.

▶ An abandoned whaling station on the shore of South Georgia, a barren island in the extreme southern Atlantic.

had the rich calving grounds of the gray whale, which had only been discovered off the coast of Baja California in 1853. These had kept the Pacific whalers busy for a mere 45 years until the depleted stocks were no longer worth the effort of pursuit.

As the turn of the century arrived, whalers were shifting their attention to the southern oceans and the last unexploited stocks of whales. Whaling stations were established on the South Georgia and Falkland islands as sites where whales could be butchered and processed close to the fishing grounds. Once freed of the time-consuming obligation to process their catch, the whalers could concentrate on catching whales and towing them back to the stations. Many thousands of blue whales were processed each year by the British and Norwegians, who maintained a monopoly on the stations. By 1908 the British authorities, worried about depleting the blue whale stocks, had imposed restrictions on the harvesting of females with calves and had placed a tax on every barrel of whale oil produced on the islands. Resentful of this interference, the Norwegians developed floating whale-processing stations to circumvent the restrictions. These eventually developed into giant factory ships, which were serviced by a number of smaller catching vessels.

With the invention of the factory ships, the southern oceans became accessible for whaling to any nation. In the years up to World War II, first blue, then fin, whales were hunted. The war put a temporary stop to the decline of blue and fin whales, and their numbers recovered slightly. However, by the 1960s the operations of the continually growing whaling fleet

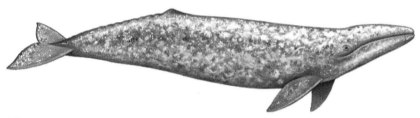

The "youngest" whale species, the grey whale has been around for 100,000 years, and is the only whale species to feed off the ocean bottom.

Probably the most common whale, the sperm whale numbers between about 500,000 and 2 million. The fatty material in its huge forehead is the source of abundant and high-quality oil.

The fin whale is the second largest whale. Like its relative the blue whale, it has a streamlined shape that gives it speed of movement.

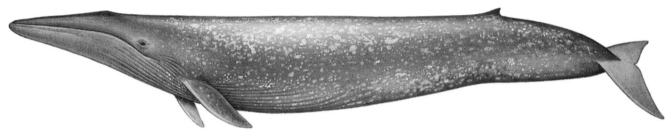

The blue whale is the world's largest living animal, and its moans are the loudest sounds made by any living animal.

▶ A record of the whale catch removed from the southern oceans between 1920 and 1980 shows that as each species became depleted, another took its place.

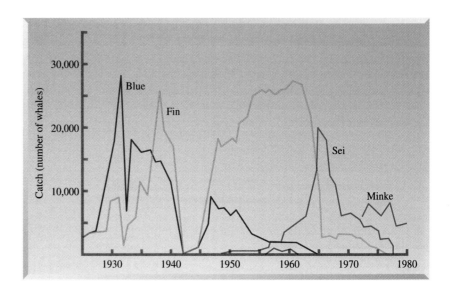

had depleted the fin and blue populations to such low levels that the whalers switched their attention first to the sei whale, and finally to the minke.

In 1930 alone, 30,000 whales had been caught; the resulting glut in the whale oil market made it uneconomical for ships to put to sea. In response, for the first time quotas were set on the numbers of whales that could be caught in the southern oceans, beginning in 1932. In 1946, a formal system of quotas was set in place by the newly formed International Whaling Commission, a consortium of most whaling nations. The quota system set a limit on the total weight of whales caught in a single year and proposed that the 110 barrels of oil produced by one of the huge blue whales was equivalent to the oil produced by 2 fin, 2.5 humpback, or 6 sei whales. Although well intentioned, the system had several drawbacks: rather than alloting each of the 17 whaling countries a quota, it stipulated that each whaling season could only remain open until the year's quota of whales was caught. An intense competition flared up between fishing fleets to catch whales early in the season, and the pressure on the more valuable blue whales only increased. Between 1946 and 1951, the number of boats doubled from 15 factory boats serviced by 129 catcher boats to 19 factory boats serviced by 263 catchers. The whaling season contracted from 112 to 64 days, and yet the stocks of whales continued to decline.

The International Whaling Commission finally established national quotas in 1961; from then on each country's whaling fleet could take only a fixed number of blue whales in any year, and the season expanded back out to 115 days. However, the total catch continued to decline, from 2.5 million tonnes (metric tons) in 1930 to less than 200,000 tonnes in 1979. That decline was of course the sign of a rapidly dwindling whale population. Eventually most whaling became economically unrewarding. The economic pressure and the public outcry against the industry in the 1970s and 1980s led to the almost complete cessation of whaling, although Japan and Norway continue to take whales for "scientific" purposes and several Inuit groups catch whales using traditional techniques.

This pattern of initially impressive yields followed later by a marked decline characterizes many of the populations that have been exploited by humans. Wild birds and spotted cats are just two examples that we will examine before looking more closely at the reasons that the pattern is so ubiquitous.

Pet Birds and Second-hand Coats

The lives of many people are enriched by the companionship of their pet dogs, cats, and birds. Dogs and cats have, of course, been domesticated for over five thousand years, and even about 85 percent of the birds sold as pets are bred domestically by bird breeders. Yet the pet trade still takes a large number of birds from the wild. The United States is the world's largest importer of birds: around 800,000 birds are brought into the country each year, and of these a quarter of a million are parrots. The declared value of these birds is around $15 million; the parrots alone are valued at $10 million. Unfortunately, between 10 and 40 percent of the birds captured die in transit

The world trade in birds is estimated to handle at least 3.5 million birds a year; parrots, birds of prey, and songbirds make up the bulk of the trade. The main bird-exporting countries—located in Africa, Asia, and Latin America—have seen wild populations of birds vanish, not just in trade to foreign nations but also to significant local markets. Many species have disappeared from areas where they were once abundant; at least two species, the Bali mynah and spix macaw, have been brought to the brink of extinction. There is not even the consolation that the local peoples gain economically: most of the profits from this trade go to the middlemen, not to people in the local community where the birds are captured.

Like the whaling industry, the bird trade has shifted from species to species: when one species becomes too rare to exploit profitably, a new species takes its place. The same cycle of exploit, extirpate, and abandon has characterized the fur industry as well.

Until recently, fur coats were regarded as the height of luxury, and the spotted coats of wild cats, especially, were an item to be coveted. Unfortunately, significant numbers of skins are required to make each garment—

► On a roadside in Mexico, a man peddles parrots captured from the wild. About 400,000 live birds were imported into the United States in 1991, a drop of about 50 percent since the early 1980s.

most coats incorporate the skins from 10 to 30 dead cats. In response to the combination of this arithmetic and high demand, the market for small cat skins from Latin America boomed in the 1960s and 1970s, with half a million skins reaching the market in each year. In the early 1970s the declining numbers of jaguar and ocelot led to a ban on their export to the United States. As a result, the majority of the exports moved to western Europe, particularly West Germany, and there was a switch to Geoffroy's cat as the main fur species.

By the late 1970s exports from South American countries had dropped to a quarter of a million cat skins a year. As overharvesting continued and the public became aware of the decline in cat numbers, the demand for fur eventually diminished, and by the mid-1980s exports from South America had dropped to almost zero. At the same time, however, the market shifted to the Far East, as leopard cats from China and the Philippines became sought after. In Europe, aggressive advertizing by antifur campaigns made the wearing of fur socially unacceptable, and the price of an ocelot coat dropped from around $10,000 in 1984 to less than $1000 in 1986.

The international trade in live birds and dead cats suggests that the profits to be made from exploiting wild animals are significant, and indeed the total world trade in wildlife exceeds $10 billion annually. However, the majority of cat and bird species that were once exploited for trade are now listed by CITES as threatened or endangered. Certainly, neither birds nor cats seem to have been exploited on anything like a sustainable basis, and the demand for their products usually has led to rapid declines in the size of the wild populations. The temptation is strong to exploit wild populations without restraint because nobody owns wild cats and birds until they have been captured. The exploiter thinks that if he or she does not take these birds (or cats or whales), someone else will. Unfortunately, similar arguments apply to the history of fisheries, forestry, and other exploitive industries.

The economist Garrett Hardin originally pointed out in 1968 that shared resources are likely to be subject to overexploitation. He envisioned a pasture, or commons, owned by a group of villagers who can each graze their cattle there. Although it is in the interests of everyone in the village not to overgraze the pasture, each individual villager is likely to see his own earnings increase from having more cattle grazing on the common land. Hardin suggested that the asymmetrical conflict between individual benefits and group costs would inevitably lead to overgrazing as each villager tried to increase his earnings at what appears as only a small decrement to each of the other villagers. This conflict between individual gains and community costs is known as "the tragedy of the commons." Only in cases where strong tribal recriminations prevent any individual from taking more than his or her share can the inevitable tragedy of overexploitation be avoided.

In fact, there are almost no examples of sustained human use of natural resources that have not led to overexploitation and the near extinction of the exploited species. Don Ludwig, Ray Hilborn, and Carl Waters of the University of British Columbia in Vancouver have surveyed the history of natural resource exploitation and have pointed out a remarkable consistency in the trends underlying almost all of the well-studied examples. They identify four common features that epitomize the failure to manage natural resources in a sustainable fashion: (1) wealth or the prospect of wealth generates social and political power that is used to promote unlimited exploitation; (2) scientific understanding of exploited systems is hampered by a lack of controls and replicates; (3) the complexity of the underlying biological system precludes a simple reductionist approach to management, and instead optimum levels of exploitation must be learned by trial and error; and (4) large levels of natural variability mask the effects of overexploitation, so that it is often not detected until too late.

The Worm in the Budget—Why Sustainable Development Usually Fails

In most cases the exploitation of a natural resource follows a consistent sequence of events, well illustrated by the history of whaling. Initially, the substantial profits made at the beginning encourage more people to enter the industry. Increasing competition then stimulates the development of new technology that makes possible increased yields; the stock inevitably continues to decline. Eventually, quota systems have to be applied to prevent the industry from driving the stock to extinction, and these lead to further competition, diminished earnings, and often to government subsidies. For example, many nations subsidize their fishermen to continue catching fish once stocks become depleted. The fishermen and their families naturally vote for the political party that promises them the biggest subsidy. So once the subsidies are in place, it is very hard for a government to remove them without losing votes at the next election. Statistics from the fish and agricultural organizations of the United Nations show that in 1992 the world's commercial fishermen spent $124 billion catching $70 billion worth of fish! The shortfall is mainly made up by government subsidies. As we have seen in the previous chapter, similar subsidies encourage overlogging of forests and overgrazing on public lands.

Whether whales, elephants, spotted cats, parrots, or forests are the natural resource at stake, a number of factors seem to have discouraged their sustainable use. In each case, the initial development of a market led to an increasing economic demand for the product. Because most ecological populations take years or decades to grow, demand rapidly outpaced supply and the price rose as the resource became scarce. Rising prices only increased the incentive to further exploit the stock, until the population could no longer replenish itself. In many cases government subsidies kept people employed in industries long after they had ceased to be viable and provided the final push that caused the resource to collapse. All of these patterns imply that market forces (the "invisible hand" of Adam Smith) have been woefully inadequate in ensuring that supply can always meet demand. Whales, parrots, ocelots, and mahogany trees are not like personal computers or refrigerators—an increased demand for them does not lead to a huge increase in their natural reproductive rate. Quite the contrary: a big dip in their numbers reduces the rate at which they can produce an exploitable harvest.

The competition for unprotected resources is further compounded by another important economic factor: to many people in business, purchasing a tract of forest for its timber, or a deep-sea fishing vessel, constitutes a substantial investment. If they wish to receive a continuous source of income from that investment, they can harvest the resource at some low level and live off the annual profits. However, if they wish to maximize their total profit, they may be better off exploiting the stock to extinction and investing the short-term profit in economic ventures that give a more substantial annual return on their investment. If annual profit is the only criterion we use to determine which is the better strategy, then, as explained below, the level of exploitation will be entirely dependent upon the

In the Azores, a whale is flensed for its blubber.

population growth rate of the species we are exploiting and the growth of money if invested elsewhere. This process is termed *capitalization*. It was first described in a forceful and insightful analysis of the whaling industry by Colin Clark at the University of British Columbia.

Let us consider this argument using rough figures derived from postwar whaling in the southern oceans. Suppose there were 200,000 blue whales in the Antarctic, and harvesting at a sustained level we could remove 5 percent of these whales yearly. If we assume that each blue whale is worth $7000, the total annual sustained yield from the harvest is $70 million a year. Alternatively, let us assume we could catch all 200,000 whales in one year and obtain a lump sum of $140 billion. If this amount is invested at a rate of 10 percent per year, the annual income is $140 million—twice that obtained from sustained use of the resource. Obviously, this example is simplistic—it serves mainly to illustrate the point that capitalization can encourage overexploitation, particularly when the exploited population grows slowly. In a more realistic world, it would be increasingly expensive to harvest the last few whales; on the other hand, 5 percent may be an optimistic growth rate for blue whales, and many shrewd investors would expect a higher return than 10 percent.

If the growth rate of the exploited population is substantially less than the growth rate of the money earned by its exploitation, it is always economically sensible to overexploit the stock and invest elsewhere. If you are harvesting whales, parrots, or stock from most fisheries, the growth rate of these populations is so slow that you are financially better off overexploiting. Furthermore, even if you have managed a resource such as a forest to be sustainable, your capital is tied up in the stock. Thus you may not have large amounts of disposable income and are susceptible to competitive buy outs by organizations that have impressive amounts of ready cash but probably less commitment to preserving the forest. The incentive to overexploit and the vulnerability of more conservative managers have contributed to the destruction of forests throughout the world in the years since the end of World War II. Much of the overexploitation of forests in the United States, the Far East, and South America has been carried out by well-funded development companies that have competitively taken over from older companies with limited assets.

Capitalization is a major obstacle to sustainable management, unless overexploitation ultimately reduces a business owner's ability to benefit from the wealth he or she has accumulated. In later chapters on sustainable development and global change I will argue that the overexploitation of forests and oceans is beginning to have an impact on the planet's welfare that may considerably reduce the benefit that these individuals will receive from their wealth.

The world's fish catch dramatically illustrates the consequences of capitalization and government subsidy. The total weight of fish caught worldwide each year has increased almost fivefold since 1950, mostly because fisherman have abandoned their overexploited home waters for less heavily exploited areas, particularly in the tropics. Unfortunately, the size of the world's fishing fleet has grown twice as fast as the annual catch, so fishermen are taking smaller-sized fish than they would have accepted before, and a wider range of species. All this activity has recently produced the inevitable decline in the total world fish catch, which reflects the dwindling of the world's marine fish stocks. This is particularly bad news for the poorest two-thirds of the world's population, who receive about 40 percent of their protein from fish. Recently, D. Pauly and V. Christensen, of the International Center for Living Aquatic Resources Management in Manila, have examined world fish catches for the years 1988 through 1991 (94.3 million tonnes of landed fish and 27 million tonnes of bycatch, which are caught fish that are thrown away as unmarketable). Their analysis suggests that fishermen removed an amount equivalent to 24 to 35 percent of the fish produced each year in fresh water and near the shore. Although only 8 percent of the fish produced annually in the deep oceans were removed, this figure is four times as large as previous estimates. Harvesting at this level will continue to deplete the stock of fish in the world's oceans.

With so many factors encouraging overexploitation, legislation has often seemed to be the best way of ensuring sustainable use. Unfortunately, most biological resources are exploited so rapidly that the needed legislation is proposed only when the resource is almost beyond recovery and at a time when its rarity tends to increase its value as a commodity. Yet even

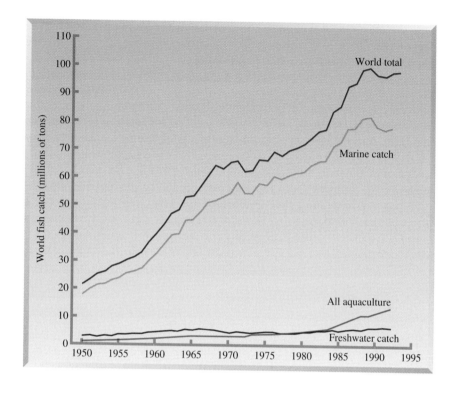

◀ The annual world fish catch has quadrupled in only fifty years, and by far the largest proportion is taken from the open ocean.

when it is not too late to save a species, preventing overexploitation presents many difficulties: for example, we have to be able to compute how many individuals may be safely removed from a population. Some harvesting models that were originally developed for fish illustrate how difficult setting sustainable quotas can be.

The Dynamics of Harvested Populations

Anyone hoping to produce some form of sustainable product from wild animals or plants would probably prefer to take an *optimum* harvest, one that takes as many individuals as possible without overexploiting the population. The problem obviously is to determine how many individuals may be removed from a population without endangering its survival; the solution requires an understanding of the dynamics of populations that suffer regular losses as individuals are taken away for human use. We need to know how the unexploited section of the population will respond to removal of these individuals: Will the population decline? By how much? Can the population eventually recover?

An immediate consequence of harvesting a population is a decrease in the size of that population. However, most populations produce an excess of offspring that must compete with each other for food. As the available food resources are finite, some offspring fail to obtain enough to eat and these die. Alternatively, the females in a population must consume a fixed amount of food in order to conceive, so when food becomes limited, the birth rate declines and the population stabilizes at a density roughly determined by the available food. In many cases, reducing the number of adults alive now by harvesting them will allow the birth rate to increase or a larger proportion of offspring to survive. The new births or the decline in juvenile deaths may offset the loss of some individuals from exploitation.

▼ The beginnings of Cannery Row's sardine industry. This photograph, taken in 1911, shows sardines being left out to dry at the Pacific Fish Company's docks.

Ecologists call such compensatory processes density-dependent mechanisms; much of harvesting theory is involved with their application to understanding the relationship between *stock*, the size of the population at any time, and *recruitment*, the rate at which new individuals enter the population at any population density. In an ideal economic world where supply met demand, the removal of some individuals would be perfectly compensated for by an increase in recruitment. However, animal and plant populations contain intrinsic developmental time delays, since it takes weeks, months, and often years for a newborn individual to develop into a reproducing adult. Furthermore, the birth and death rates of the population will vary with fluctuations in weather, in the number of predators, or in the number of species that compete for the same food resources. The supply of new individuals does not respond to demand as it would in a manufacturing industry.

Two interlinked examples show how quickly a population can crash in numbers when recruitment doesn't compensate for demand. The California sardine fishery made famous by John Steinbeck in his novel *Cannery Row* was established to process the 60,000 tonnes of California sardines that landed each year at Monterey Bay in the first half of this century. Toward the end of the 1940s the catches became increasingly erratic, and by the mid-1950s the fishery entered a sharp decline. Eventually, the California sardine industry collapsed and Cannery Row was shut down. The stock failed to recover once fishing stopped, and it was suspected that something, perhaps the decline of the fishery itself, had instigated some long-term change in conditions unfavorable to the sardines. One sign that such a change in conditions had taken place was that anchovy, formerly rare, were now highly abundant throughout the eastern Pacific. The increase in anchovy was capitalized on by the Peruvians, who developed a highly successful fishing industry that exploited anchovies throughout the 1960s. However, this fishery too collapsed in the 1970s mainly due to overfishing—it was estimated that in some years nearly half the stock was removed.

Harvesting and Fisheries

It is never a trivial exercise to determine the relationship between stock and recruitment for a biological population; to do so we usually need data collected over a number of years, for a range of population densities. Initially let us consider an ideal case where a population has been exploited at a range of densities and we know the relationship between stock and recruitment. For any population size, the rate of population growth, or recruitment, is the product of the number of females in the population and numbers of surviving offspring produced by each female. At very low population sizes, as long as every female in the population is able to find a mate, then the number of surviving offspring produced per female will be at a maximum. However, the total number of individuals added to the population each year is still low, as there are only a few individuals actually producing offspring. Although the population may increase at its most rapid rate, the total annual increase in numbers is not substantial.

Almost the opposite effect takes place at high population densities. In this case, the habitat may contain all the animals it can support, and competition for resources is intense. Although the females in the population are many in number, they give birth less often; nor are infants and juveniles as likely to survive. At these high densities, most individuals in the population leave only enough offspring to replace themselves. The number of births may go up when animals are removed, but competition for resources remains intense and many offspring still do not obtain the food they need to survive. As a result, many individuals lost to harvesting may not be replaced, and the population declines to lower density.

In contrast, at intermediate densities, the rate at which individual females produce surviving offspring is still high and yet their numbers are also significant: many new individuals may be recruited to the population, while competition for resources may be low. Thus it is usually at intermediate densities that we obtain a maximum yield, although the optimum density will depend on the way the species is exploited and on aspects of its biology such as its social organization and the resources it uses.

The biologists who study fisheries refer to the level of exploitation that maximizes yield as the maximum sustainable yield, or MSY level. For many years, achieving the MSY level was the underlying goal of the management policy guiding the exploitation of many fish stocks. There is now widespread agreement that this was a dangerously oversimple approach to managing fish populations. Ensuring that fishermen do not exceed the MSY level is tremendously difficult, even after the level is known, but estimating the actual level of exploitation that produces a sustainable yield is itself a formidable task.

Effort and Yield: Stock Recruitment Curves

Because it is often impossible to estimate the sizes of oceanic fish populations, for most fish stocks we know only the number of individuals removed from a population—the yield. The actual size of the stock that is being exploited has to be obtained by indirect means. One alternative index of population size is fishing intensity, measured as catch per unit effort, with effort usually defined as the number of days a boat spent fishing. (In some heavily exploited fisheries, such as the Pacific salmon, the effort is now measured in boat hours, because the fishery is only open for a couple of hours at a time.) If we plot catch against fishing effort, we may obtain some understanding of the relationship between stock and yield. Although some fisheries give fairly good relationships between fishing effort and volume of fish caught, others are considerably less well behaved because of the influence of environmental factors, such as water temperature, on recruitment. Although biologists can obtain some estimates of how these factors affect the fishery, they usually require data collected for many years over a range of population densities.

Once we have an idea of the stock, or actual numbers of fish, the next step is to figure out at what size stock the recruitment is at a maximum. There is no simple (or even complicated) formula that will give us the answer. Instead, fishery managers have to learn the optimum stock size, and the optimum harvest, by trial and error, through the efforts of the fishermen themselves.

In an "unmanaged" fishery, the population of fish will tend to fluctuate, as the water temperature or other conditions change and as the intensity of fishing varies. Managers can look at the different stock sizes, match them up with the fishing effort, and see how much the stock size rose or fell in response. Eventually some stock size and catch will reveal itself as the possible optimum size and harvest. If fishermen are not harvesting the fishery at a sufficiently wide range of densities, then we may never fully understand the relationship between stock and recruitment. It is ironic that a fishery that is tightly managed to provide consistent yield at some constant effort will provide very little information on the relationship between stock and recruitment. In contrast, a slackly managed fishery, which is being harvested at a range of densities and yields, will provide more information with which to manage the fishery!

Once the MSY is established, the total yield taken from the ocean has to be regulated. The usual method is to ensure that only fish larger than certain age classes are taken, perhaps through regulations on net sizes or the use of lobster gauges. These techniques prevent immature individuals from being removed before they have had a chance to reproduce. Although such regulations are widely in force, they do not seem to have produced the desired result. In a survey of managed fisheries for the U.N. Fisheries and Agriculture Organization, John Gulland pointed out that out of the 86 fisheries for which long-term data were available, at least 78 had been overexploited and were no longer viable. Was this overexploitation simply an accidental result of the lack of key information about the stock size, was it human greed disguised as the tragedy of the commons, or does it represent some more complex feature of biological exploitation? The answers are still uncertain.

For further insight into the relationship between population size and exploitation, consider the harvesting of elephants for their tusks, the mainstay of the ivory-carving industry. Being the largest living land mammals, elephants can be readily counted from either the air or the ground. As independent information can thus be obtained on stock sizes, it should be possible to exploit these animals in a relatively efficient manner. Yet the manner of their exploitation was far otherwise in many countries until elephants became the cause celèbre of the conservation movement in the 1980s, when a variety of sources indicated that their numbers were declining precipitously.

A History of the Ivory Trade

It is not known when Indians began to carve ivory, but there are wonderful ivory carvings that date from the time of the Harappan culture (2500 to 1700 B.C.). The Indians of that time carved the material into combs, figurines, and jewelry. Ivory carving remained a high art form practiced by a few skilled carvers, who were supported by wealthy artisans, until the British invasion of India in the seventeenth century. The arrival of the British created a larger market for smaller ivory pieces—in particular, for chessmen and piano keys. The sharper tools that became available at this

▶ An ivory carver practices his craft in Hong Kong.

time made possible the rapid development of a form of mass production to meet the new demands. During the late seventeenth and early eighteenth centuries, carving ivory declined from an art to a craft.

As the demand for ivory grew, supplies from Indian elephants were supplemented with the softer, larger ivory obtained by importing tusks from Africa. The first major wave of poaching to hit Africa's elephants began in the late eighteenth century, and poaching continued to take a heavy toll on African elephants until World War I introduced a lull in the demand for ivory that lasted a few decades. After World War II, the center of the ivory-carving industry moved from India to Hong Kong, where the trade tended to concentrate almost entirely on ivory from Africa's elephants.

The international financial crises of the 1970s and the development of Japan as a major economic power increased the demand for ivory jewelry and the ivory signature seals that are used in Japan to seal letters and messages. The price of ivory soared. As a result, in the 1970s and 1980s an epidemic of poaching spread sequentially through the countries of Central and East Africa. Ironically, several major conservation organizations continued to support the trade in ivory as a viable means of conserving elephants, partly because they were receiving donations from the ivory trade and partly to retain the support of government officials who were financially involved in the trade.

By the mid-1980s more than half a million elephants had disappeared from East and Central Africa. Indeed, the decline of elephants in many African countries almost exactly matches the rate of ivory export. The Kenyan government vividly drew the world's attention to the plight of the African elephant on July 18, 1979, by burning a 12-ton mound of confiscated ivory in Nairobi National Park. This $5.5 million bonfire was intended as a public statement that Kenya and several other East African countries wanted CITES to grant the African elephant full status as an endangered species and to stop all trade in ivory before elephants disappeared entirely from much of their range in that part of the continent. The call for a ban was controversial, as several southern African countries claimed to use the revenue from ivory to pay for conservation schemes. Unfortunately, these countries ignored the fact that they were benefitting from other countries' attempts to ban ivory trading, since this revenue was obtained from ivory whose price was considerably inflated by its shortage.

▲ Ivory worth $5.5 million goes up in smoke, part of the Kenyan government's campaign against elephant poaching. Almost 20,000 elephants were killed in a single year in Kenya during the peak of the poaching epidemic in the early 1970s.

The Demography of African Elephants

At present, killing elephants for their ivory is simply forbidden. But suppose that the elephant population recovers sufficiently that some harvesting of the population becomes permissable. What would be the best strategy for achieving sustainable use? Using the data gathered in a variety of East African ecosystems, we are able to construct a population model from which we can determine the relationship between population size (or stock), exploitation level, and persistence for African populations.

Let us assume that elephant populations can be exploited in two different ways. One approach is to cull animals in order to control population density: the method calls for whole family groups to be removed from the population by shooting. This is the method that is employed in Zimbabwe, Botswana, and South Africa, where farmers are allowed to kill elephants to limit the damage the animals wreak on crops. The alternative is to cull elephants in order to maximize the yield of ivory; here only elephants with large tusks are removed from the population. These two approaches have a different impact on the size of the population and the yield of ivory.

The graphs on this page illustrate the case where random groups of females are culled in order to control population density. Maximum sustainable yield is obtained when on average 1.0 percent of the females are removed in a year. When this percentage is increased to 6 percent, the population declines rapidly to extinction. Notice that as the average harvesting level increases, the time it takes to drive the population to extinction decreases and the time it takes to reduce the population in size by 50 percent also decreases. Obviously, a decrease in total population size of

▼ Left: When elephants are removed at random from a population, the ivory yield is highest at an annual culling rate (given by the numbers above the curve) of 1.0 percent of the population. At this rate of culling, the population will maintain a size about half as large as a population that isn't culled. Right: Any culling level above 2 percent of the population per year will rapidly lower the elephant population to below 50 percent of its original size; a culling level above 6 percent will lead to extinction.

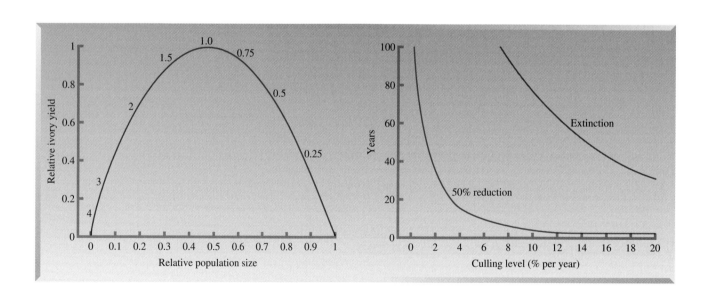

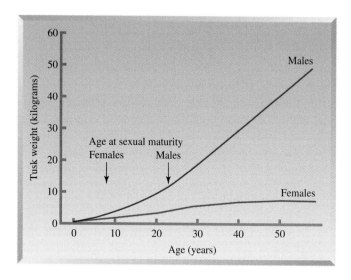

◀ The tusks of male elephants continue to grow larger and heavier with age, while those of female elephants level off at a much smaller size.

50 percent will correspond to a decrease in ivory yield of 50 percent if the population is being exploited at random.

Rough estimates suggest that by the late 1970s approximately 17 percent of Kenya's elephants outside the parks were being killed each year, and 8 percent inside the parks. These figures suggest that Kenya's elephants would have been driven to extinction in 30 to 40 years, or toward the second decade of the next century. It was for roughly this reason that Kenya called for a halt to the trade in ivory.

In contrast, we can compare what happens if elephants are selectively harvested to maximize ivory yields. To harvest an elephant population for this purpose, we need to know the tusk sizes of animals of different ages and sexes in a population. The relationship between tusk size and age is illustrated in the graph on this page; the tusks of male elephants are consistently much larger than those of females, and the disparity of tusk sizes increases with age. If we were selectively harvesting elephants for tusk sizes, we would start by removing the largest males and switch to females only once the tusk size of the remaining males was approximately equal to that of females. Notice that the point at which we switch to females occurs more or less after we have exhausted all the sexually active males in the population. What happens to the relationship between ivory yield and population size?

The new stock-yield relationship is illustrated in the left graph on the facing page; the relationship between stock and yield is now considerably skewed toward the right, with the maximum sustainable yield obtained when the population is at around 95 percent of its maximum density.

Under these conditions the ivory yield is more than 10 times higher than the yield obtained when elephants are killed at random. Far more ivory is harvested, yet many fewer elephants are killed.

Removing only animals with the largest tusks has a very dramatic effect on the sex ratio in the population. As the population shrinks, it first loses most of the sexually active males; once the total population falls below 95 percent of its maximum density, we are essentially removing mature females and immature males. The relationship between tusk size entering the trade and the yield of ivory follows no easily grasped pattern, but the maximum yields are achieved by harvesting tusks in excess of 20 kilograms. Yields of ivory—and the size of the surviving elephant population—diminish rapidly as we move below 10 kilograms.

Unfortunately, the tusks entering the trade from many East African countries were between 3 and 5 kilograms in size during the ivory poaching epidemic of the 1980s. Populations surveyed in different parts of East Africa by Joyce Poole of Kenya Wildlife Service exhibited the extreme sex ratios illustrated in the graph on the right below. Poaching must have not only pushed elephants to low population levels, but also strongly disrupted their social system. In contrast, southern African countries such as Zimbabwe and Botswana had not yet experienced the epidemic of poaching, and the tusks exported from these countries reflected the culling of groups of females to control a population. No country was managing its elephants in a way that would maximize ivory yield.

We can use the data on tusks entering the trade to estimate the recovery times for populations in different countries. We determine the ages

▼ Left: When elephants are culled by removing the individuals with the largest tusks, the culling rate that yields the greatest weight of ivory will leave the population at 95 percent of its maximum density, a much better result than the 50 percent left after random culling. Right: Males are removed first when elephants with the largest tusks are culled. Thus the more animals are culled, and the lower the population declines, the higher the ratio of mature females to mature males.

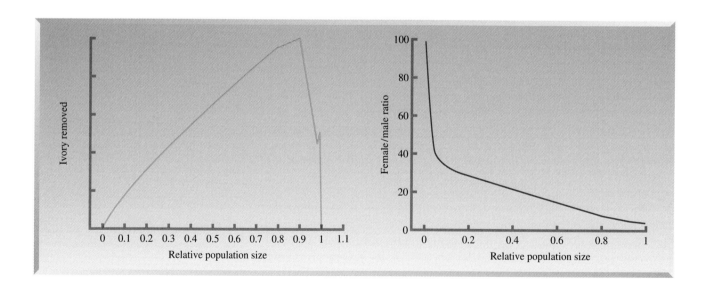

of the elephants from whom the tusks were removed, enter these ages into the model as the oldest possible age classes, and run the model forward. These calculations suggest that it will take between 30 and 50 years for most of the heavily poached populations to regain half their average size before poaching began and in excess of 100 years for complete recovery.

Obviously, whether or not to manage elephant populations using profits from the sale of ivory will continue to be a controversy. Unfortunately, very little of the debate has looked at how the size of the elephant population shifts in response to the size of the average tusk harvested. The analysis described here suggests that if maximizing ivory yield and raising revenue to pay for conservation is the goal of management schemes, then random culling is obviously not the best way to harvest elephants.

Hopefully, in areas where elephants interfere with human activities, alternative, more ethical, ways of controlling their populations will be developed. Controlling fertility, for example, might be a way to regulate very dense elephant populations in areas where they are causing damage. New ways of managing elephant populations may allow the trade in ivory to be reopened some time in the future, perhaps in twenty to fifty years. It is essential for the long-term viability of elephants that the profits of this trade go back into wildlife conservation, rather than into the pockets of a few illegal dealers and profiteers.

Strategies for Ending Overexploitation

Perhaps the main reason that the sustained use of wild animal and plant resources has failed so miserably is that so much of the trade has been undertaken illegally. In many cases, government officials have either looked the other way or, all too often, illicitly profited from the trade. At the height of the ivory trade boom in the 1980s, only 5 percent of the ivory entering the market was obtained legally; the vase majority was either poached (80 percent) or recovered from poachers (15 percent). The hand of government corruption is clearly visible in the CITES export figures for many countries. Those export figures often record a shipment to another country as a factor of 10 smaller than it appears in the import figures of the receiving country. One country with a population of less than 20 elephants regularly exported ten thousand tusks (tusks that had been illegally poached in other countries were smuggled into the country to obtain export licenses). It would seem that in these instances bribery and corruption form the dominant fingers of Adam Smith's invisible hand.

Studies suggest there are three major ways to prevent corruption: (1) increase the perceived probability or severity of punishment; (2) decrease the profit from the crime; (3) decrease the incentive for the crime by improving wages elsewhere. An economic analysis of rhino and elephant poaching in Zambia, by E. J. Milner Gulland and Nigel Leader-Williams, comprehensively illustrated that poaching was profitable. Although antipoaching patrols were able to apprehend poachers, the jail sentences were too slight to deter the poachers from returning to this career after their release. Moreover, running antipoaching patrols is expensive; it costs more than $200 per square kilometer to successfully protect game in African parks. Although antipoaching patrols could recover up to 40 percent of their costs of operation by selling recovered rhino horn and ivory, the dead elephants and rhinos cannot easily be replaced. Attempts to reduce the profit margin of poachers by removing the horns from rhinos have proved unsuccessful; around 9 percent of the animals die under the anaesthetic, and the dehorned animals are unable to defend themselves in fights with other rhinos or against predators.

A more successful strategy, proved effective for both elephants and spotted cats, has been to weaken the market by making ivory and fur coats socially unacceptable. Witty and outrageous advertising campaigns will continue to influence public opinion about other overexploited and endangered species; in the age of the soundbite they provide a cost-effective way of manipulating the markets for wildlife resources.

Using stochastic models like those in Chapter 3, Russ Lande has shown that extinction is the inevitable result of harvesting any population when it is below the natural density it would settle to in the absence of ex-

▶ This black rhinoceros, photographed in Zambia, was killed by poachers for its horn.

ploitation—often termed its carrying capacity. This conclusion comple-
ments Colin Clark's result that species whose population growth rates are
less than the growth rate of money are also doomed to overexploitation.
There is a general trend for population growth rate to decline with increasing body size. Once we begin hunting animal species whose adult
mass is larger than around 100 kilograms, we will almost inevitably end up
overexploiting them, because their population growth rates are too slow to
compensate for any substantial harvesting. Not surprisingly, we have been
most efficient at exploiting smaller animal and plant species, including animals such as goats and sheep, and cereal crops such as rice, wheat, and
corn.

Reducing the consumptive use of wildlife does not necessarily lead to
reduced revenue for a country that has previously exported, say, cat skins
and elephant tusks. In each of the cases described above, human distaste for
treating wildlife as commodities has been followed by a rising interest in
wildlife tourism. The elephants of Amboseli National Park in Kenya annually raise many times their value in ivory from tourists who pay to visit the
park. Nearly 20,000 Japanese paid $5.4 million to watch whales off Japan
in 1992, while in the United States the total revenue from whale watching
was $130 million. Whale watching has become as profitable, and is easily
more sustainable, than whaling was at the start of the century. In many
countries, furthermore, tourism will provide a source of employment for a
great many people, including those who might otherwise become poachers.

Governments could encourage conservation by changing how they
raise and spend money—for example, by cutting off subsidies to fisherman. The European Union, in one egregious example, increased fishing
support from $80 million in 1983 to $580 million in 1990 at a time when
all stocks in European waters were declining. Poor countries could tax fisherman from wealthy countries for fishing in their waters. The Falkland Islands has introduced taxes of 28 percent of the value of the catch for fish
caught in its territorial waters. Foreign fishermen were incensed, but still
paid; the island's gross domestic product quadrupled and its fish stocks
stopped declining.

Measures like these could save many species that are at risk. But when
a species' population has dropped low enough, so that perhaps only 10 or
20 individuals are left, simply putting a stop to their exploitation isn't
enough to prevent their extinction. In these cases, biologists must consider
the more drastic steps detailed in the next chapter.

6

Species in Captivity

A female snow leopard creeps forward across a cold rock surface, her eyes fixed on a Temminck's tragopan, a magnificently colored species of pheasant confined to the higher slopes of the Himalayas. Distant gunfire disturbs the tragopan and it flies to a new perch. The snow leopard rolls over and goes back to nursing her three cubs. Eight onlookers click their cameras, two children demand to be taken to the washroom, and three others remain transfixed by the scene behind the protective glass in the small cave in front of them. The million annual visitors to the Bronx zoo in New York see scenes such as this on almost an hourly basis, yet only a handful of people have ever seen a snow leopard in the wild.

The Himalayan highlands exhibit at the Bronx zoo illustrates zoos at their best: the animals live in a natural setting, and they behave similarly to their wild counterparts. The snow leopards and tragopans are part of successful captive breeding programs designed to ensure that zoos will always have healthy populations of these animals for visitors to view. The whole exhibit highlights

◀ Snow leopard in the Bronx Zoo, New York. Over a million people see snow leopards in captivity in the Bronx. Less than five people have ever photographed them in the wild.

▶ Zoos have come a long way since the early years of the century! Contrast this "snow leopard exhibit" of 1906 with the one illustrated in the chapter opener.

snow leopards as creatures that have adapted to life in an extremely harsh mountain climate. On a more subtle level, the graphics at the edge of the exhibit emphasize how trivial human whims, epitomized by the fur and fashion trade, could quickly drive snow leopards to extinction.

Zoological parks first became popular in the nineteenth century, although private menageries and collections have probably existed since Egyptian times. The American Zoo and Aquarium Association (AZA) estimates that in 1994 nearly 120 million people visited zoos and aquaria in the United States—more than the combined number of people attending professional football, baseball, basketball, and hockey games! These zoos and conservation centers are run by 17,700 staff, who tend to 493,620 animals, and are supported by combined general operating budgets of nearly $980 million.

Zoos, aquaria, and botanical gardens are plainly big business, and they are also a major medium for educating a wide audience about the diversity of life on earth and its present plight. But zoos and conservation parks also seek to present themselves as storehouses of biological diversity, whose contents will eventually be used to restock the world once the exploitation of natural resources has considerably slowed. The latter is a more precarious proposition, since it is not clear that zoos can provide a viable alternative to a species' natural habitat. This chapter examines the population biology and economics of captive breeding and the logistics of designing programs to reintroduce species into the wild after they have either disappeared from their natural habitat or become endangered there.

Zoos as Noah's Arks

For those who manage captive populations, the primary objective is to ensure the propagation of these populations so that zoos and botanical collections have a constant supply of individuals to exhibit without having to remove more individuals from the wild. In some cases, though, a more ambitious goal takes precedence: managers may hope to eventually reintroduce captive populations into the wild. For example, no wild individuals of the Chinese maidenhair tree *Ginkgo biloba*, or ginkgo, have ever been located. For many centuries this primitive tree species was preserved in the courtyards of temples in China. Thanks to its introduction into a variety of habitats, it is now distributed throughout the world. The ginkgo is considered to be the first species saved from extinction by human efforts to reintroduce it into the wild. As an added bonus, it is now widely used as the source of a drug that slows senile dementia—a disease that causes loss of memory as people get older.

◀ This condor chick, removed from its parents before hatching so that they would produce another egg, must now be hand-reared by its human keepers at the San Diego Zoo. The keepers employ condor-head hand puppets so that the chick learns it is a condor rather than a human being.

The California Condor

The last California condor *(Gymnogyps californianus)* known to exist in the wild was trapped on Easter Sunday of 1987 in the Central valley of California. It soon joined 28 others of its species at the San Diego and Los Angeles zoos. The bird was a young adult male that had been followed closely for a number of years. He had been observed to pair with an older female in late 1985, and his first breeding attempts in 1986 had been documented in considerable detail. The capture of that lone male brought to an end an intensive period of study focused on California condors in the wild. Now all hope for the continued persistence of the condor would depend on captive breeding.

Since 1800 California condors had been reported as year-round residents of a region extending from British Columbia south to Baja California. They were rarely seen north of California after 1895, however, and in this century, all confirmed nest sites have been located south of San Francisco Bay and north of Baja California. Throughout this century their numbers steadily declined as the human population expanded in southern California. Condors were frequently shot for sport or by farmers who feared the birds had killed lambs or calves. They preyed on birds and rodents contaminated by DDT and poisons. Autopsies revealed some birds to have died from lead poisoning—the result of feeding on carcasses filled with lead shot from hunter's guns. Simultaneously, their natural habitat was being lost as fruit farms, roads, and shopping malls spread around Los Angeles, Santa Barbara, and the Central valley. By 1978 only an estimated 25 to 30 birds remained in the wild, and even these were in danger of being occasionally shot by sportsmen and farmers.

Maintaining captive populations of California condors in the San Diego and Los Angeles zoos costs each institution about $200,000 annually. Over $20 million has been spent in the last 14 years in a large-scale effort to

If the ginkgo example is typical, it suggests that it should be easy to keep populations in captivity so that they may be reintroduced into the wild at a later date. At present several species, such as Prezwalski's horse and Pere David's deer, exist only as captive populations in zoos, since the remaining few individuals were captured when it looked as if the last population of the species no longer had any chance of surviving in the wild. Other species, like the Bali mynah and the Mauritius pink pigeon, survive

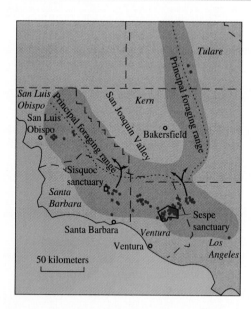

▲ Until their population dropped in the 1980s, the California condor ranged throughout the area colored in reddish brown. Their known nest sites (dots) were concentrated in and near the Sespe Condor Sanctuary in Los Padres National Park. The arrows indication the usual routes of travel of two pairs from their nesting sites to the foraging grounds.

conserve the species. Fortunately, the animals breed well in captivity. Although a pair of the birds in the wild produces at most a single egg every two years, they can be "tricked" into producing a new egg within as little as two months of the first by removing the first egg. In 1990 alone, nine pairs of captive breeding birds produced eight surviving young.

mostly in captivity, maintaining only the barest toehold in the wild. However, animals such as condors and snow leopards require considerably more husbandry than trees. If offspring born in captivity are eventually to be reintroduced into the wild, they have to learn entire repertoires of complex behaviors, which they will need in order to acquire food, avoid predators, and defend the territories that are a requirement for obtaining mates. Furthermore, as each zoo possesses only a small number of each species, their breeding has to be carefully managed to ensure that each species retains the genetic variation that allows it to adapt to changing conditions in the wild.

To assist zoos to function as "Noah's arks" for endangered species, the AZA has developed Species Survival Plans (SSPs). When first begun in 1981, the SSPs were designed to save species from extinction, although more recently they have focused on maintaining viable populations for exhibition in zoos. These plans call for the monitoring of the births and deaths in captive populations in member zoos, in part so that zoo managers can identify potential mates for captive individuals from among the populations at other institutions. The plans also provide a forum through which successful husbandry techniques can be communicated between curators and keepers at different zoos—a result has been many improvements in the design of exhibits. Currently more than 75 species are managed in SSPs, and the AZA aims to have 200 SSPs by the year 2000.

An understanding of both population genetics and population dynamics is crucial in the conservation and management of endangered species. A considerable change in scientific emphasis is noticeable as we move from managing whole communities of free-living organisms in wilderness areas, through managing populations in nature reserves, to managing captive populations in zoos and botanical gardens. While demographic considerations tend to dominate studies of free-living populations and communities, genetic and ethical considerations are more important in guiding the management of captive populations.

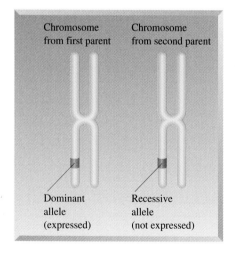

▲ An individual may inherit two forms, or alleles, of the same gene, one on each chromosome.

Breeding Species in Captivity

The goals of most captive breeding programs have changed over the last thirty years: originally the programs were operated more as informal contests to see which zoo could be first to breed a species in captivity. As the biodiversity crisis became more apparent in the early 1970s, several zoos shifted their goals to maintaining threatened and endangered species in captivity until their release into nature was possible. More recently, zoos and botanical gardens have begun to consider themselves as conservation centers whose role is to educate the public about threats to wildlife, using as an aid self-sustaining captive populations that may also provide some individuals for reintroduction into the wild.

After many years and several generations in captivity, a species may fail to survive its reintroduction into the wild unless it has managed to retain several biological attributes that are crucial for its ability to adapt to its environment. In particular, the captive breeding program must alter the genetic composition of the species and its behavioral repertoire as little as possible. Unfortunately, these two goals are sometimes very hard to meet. The behavioral repertoire will have both learned components and components that are inherited, or inate. As we move from primitive invertebrates, through fish, reptiles, and birds to mammals, a greater proportion of behavior is learned, and a lesser proportion is inate. For this reason it is essential that animals such as primates and carnivores when young spend time with experienced members of their community to develop their social behavior. Learning hunting and foraging behavior in captivity is much harder. Captive-bred golden lion tamarins have almost no opportunities to learn that snakes or poisonous frogs may be dangerous. When reintroduced into the wild, they may explore these species as potential food items—with disastrous consequences. Similarly, many zoos will come up against formidable ethical objections if they use live members of other species to train carnivores how to hunt. Yet the opportunity to practice hunting is essential for the survival of carnivores after reintroduction. When, for example, black-footed ferrets were given prairie dogs as prey, it was discovered that the ferrets' ability to chase and capture their prey was inate, but their ability to kill a prairie dog once caught had to be learned.

Its inate behavioral repertoire is of course only one of a multitude of characteristics that an animal inherits in the genetic code stored in its chromosomes. An animal's genes are stored as strands of DNA in the chromosomes located in the nucleus of each cell of the body. All animals receive two sets of chromosomes when conceived, one from each parent, and thus two sets of genes. These two sets of genes will not be completely identical, however, for the genes coding for a trait such as eye color can come in different forms, called alleles. One allele might code for, say, blue eyes and another for brown eyes. Thus each individual will have two copies of each gene, perhaps both the same allele or perhaps each one a different allele.

The alleles at the same location on each pair of chromosomes can each be either dominant or recessive. However, it is only the dominant allele, should one be present, that will be expressed; that is, only the trait coded for by the dominant allele actually shows up in the individual. A recessive allele cannot be expressed unless both alleles are recessive. Many genetic traits are coded for by more than one set of alleles, representing more than one set of genes. In other cases, the presence of more than one copy of the allele in different locations, even on different chromosomes, may increase the efficiency with which the animal (or plant) undertakes a task.

Individuals with particularly favorable combinations of alleles tend to produce more offspring in the course of their lifetimes than individuals with slightly less favorable combinations of alleles. The favorable combination of alleles will become more common in the population, making the population as a whole more "fit." This mechanism is at the heart of Charles Darwin's theory of natural selection. Although he had no idea of

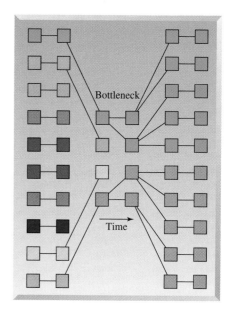

▲ Reducing a population to a few members creates a "bottleneck" that can eliminate genetic variability. The different colors indicate different genetic alleles.

the details of the underlying genetic mechanisms, he realized that natural selection could only operate if the traits present in the population showed variation. Later workers, particularly Ronald Fisher, showed that a population responds more quickly to the pressures of natural selection if the trait in question varies widely in the population.

When we take a small number of individuals out of the wild to raise in captivity, some alleles present in the original free-living population may not be represented in the captive population. In fact, many captive breeding programs are established only after the natural population has already been reduced to a very few individuals, so much of the genetic diversity of the species has already been lost. Furthermore, the rarer alleles may be lost from generation to generation by chance in a way that is analogous to the way in which demographic stochasticity leads to the loss of individuals in small populations. Thus captive populations tend to lose genetic diversity and with it their ability to respond to natural selection.

If there is any genetic variability left in the captive population, we then have to worry that captivity may induce its own selective forces that are similar to those exerted when wild animal species are domesticated. Populations in captivity may then quickly alter from their genetic constitution in the wild. For example, domestic cattle are descendants of individuals that were selected for their docility and lethargy. Aggressive animals that tend to attack each other, or their keepers, are likely to be isolated and kept from breeding. But if we selected for more docile traits in captive deer populations, we would produce a population that was much more susceptible to predators when reintroduced into the wild.

The managers of captive populations face two dilemmas: how to maintain genetic diversity without letting populations lose those characteristics that made them fit for life in the wild, and how to maintain genetic diversity in a small population that has probably already lost many alleles. Maintaining diversity is crucial if species are to be able to evolve and adapt to their surroundings when eventually reintroduced back into the wild. Genetic variability is particularly important when a population is threatened by an infectious disease. If all the individuals in a population are identical, a pathogen may be able to infect and even kill the entire population. In contrast, a host population with ample genetic variability, particularly in alleles that code for immunological efficiency, may include some individuals that are resistant to the pathogen. The infectious disease will have less impact on the population now, and even less in the future after disease-resistant individuals have been selected for.

Their small founding numbers make it likely that captive populations will be plagued with additional genetic problems. One such problem is inbreeding depression: when close relatives mate, rare and deleterious genetic defects are likely to be expressed in their offspring. We know of two types of inbreeding depression. First, rare lethal traits may be expressed

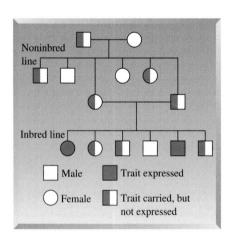

▲ Two unrelated individuals who mate (top) are unlikely to share the same recessive allele; some of their offspring may carry the trait, but they won't have inherited the two copies of the allele needed to express it. Two related individuals who mate (bottom) are far more likely to share the allele and to produce offspring that express the trait.

The Proportion of Juvenile Mortality in Noninbred and Inbred Progeny in Eight Species		
	JUVENILE MORTALITY (SAMPLE SIZE IN PARENTHESES)	
Species	Noninbred	Inbred
Degu	0.106 (66)	0.235 (51)
Elephant shrew	0.153 (144)	0.233 (43)
Pygmy hippopotamus	0.245 (184)	0.549 (52)
Cheetah	0.263 (194)	0.442 (43)
Doreas gazelle	0.280 (50)	0.595 (42)
Japanese serow	0.288 (73)	0.565 (62)
Greater galago	0.308 (146)	0.3386 (44)
Golden lion tamarin	0.501 (369)	0.634 (145)

when an individual receives a copy of a deleterious recessive gene from both its parents, as is more likely to happen when the parents are closely related individuals sharing a similar genetic makeup. Second, there is a more general decline in fitness because the offspring from matings between close relatives have less variability in their genetic makeup.

The captive populations of black-footed ferrets have suffered from the first type of inbreeding depression. Several of the captive-born females seem to carry genes that, by producing a collapsed uterus, prevent them from successfully producing young.

Similarly, the endangered Florida panther has several inbred traits, including heart defects that cause early death and undescended testicles in males. Even the cowlick and kinked tail that distinguish the subspecies are probably artifacts of inbreeding. Studies of the naturally inbred population of lions in Ngorongoro crater in Tanzania by Craig Packer and his colleagues show that sperm defects are common in inbred males and could be reducing fertility. Katherine Ralls and Jonathan Ballou of the Smithsonian Institution have demonstrated that because of inbreeding depression juvenile mortality is significantly higher in the offspring from closely related parents in a wide variety of captive populations of gazelle, deer, and other ungulates. Their studies actually underestimate the more severe consequences of inbreeding that a population would experience following reintroduction back into the wild. Captive populations are isolated from a variety of threats that could seriously deplete their numbers where predators, disease, and competitors are present.

▲ This subspecies of ibex (*Capra ibex ibex*), from Hohe Tanern National Park in Austria, became extinct after mating with other subspecies as part of a translocation project in Czechoslovakia.

Genetic studies of the fruit fly *Drosophila* suggest that there is an important interplay between the two types of inbreeding depression. In particular, the expression of deleterious recessive alleles probably accounts for only about half of the reduction in survival and fecundity that has been observed in inbred populations. In some cases a short bout of low numbers will have little effect on the amount of genetic variation, if the population can be increased rapidly to a higher population density. A population that has weathered a bout of inbreeding at low population density is said to have passed through a bottleneck. In some cases it may emerge with its fitness enhanced because inbreeding exposes deleterious recessives and allows them to be purged from the gene pool. This is precisely the method used by animal breeders to remove deleterious alleles.

Most natural populations are subdivided into geographically separated groups, or subpopulations. The subpopulation in any area may become adapted to the local environmental conditions there. When individuals from different subpopulations are brought together in captivity, new combinations of matings may be set up that would not happen in the wild, where geographic isolation keeps subpopulations from encountering one another. Although the individuals may look identical and mate with each other willingly, they may have subtle physiological adaptations that could be lost or scrambled in their offspring. This loss of adaptations peculiar to a subpopulation is often called "outcrossing depression," though actual examples are rare.

The most commonly cited case of outbreeding depression is the unsuccessful translocation of several subspecies of ibex to the Tatra mountains in Czechoslovakia, where the native subspecies had gone extinct. Ibex from Austria, *Capra ibex ibex*, were first introduced, successfully, to the Tatra mountains, where they established a small population. However, this subspecies of ibex was then supplemented with bezoars from Turkey *(C. ibex aegagrus)* and Nubian ibex from Sinai *(C. ibex nubana)*. The different subspecies mated and produced hybrid offspring. The hybrids mated early in the fall, rather than in the winter, and young were born in February—the coldest part of the year—and died. Eventually the entire population became extinct.

Computer Dating @zoo.ark

In their ability to readily identify the individual animals in a captive population, zoo managers have an advantage over the managers of free-living populations. Zoos record births, deaths, and successful matings for each captive population. Sometimes, as for the red wolf, complete genealogies are available. In the late 1970s the remaining red wolf population was pulled from its native habitat in the southeastern United States when it seemed doomed to extinction, and sent to the Port Defiance Zoo in Tacoma, Washington. The red wolf's genealogy traces each animal's ancestors back to the wild individuals that were originally brought into the zoo to found the captive population. With this kind of information stored on computer, and the ability to closely control the formation of breeding pairs, breeding programs in zoos have been able to concentrate heavily on genetic management. Although the desire to prevent inbreeding depression provided much of the initial impetus to manage these programs scientifically, it is now thought better to focus on maximizing genetic diversity.

The key to maintaining genetic diversity is to equalize the numbers of offspring produced in each generation that are descendants of the original individuals captured in the wild. That means that, as much as possible, all individuals in the population should leave the same number of offspring. Equalizing reproductive success in the captive population not only helps maintain genetic diversity, but also reduces the opportunity for selection to operate. Docile animals, say, don't have the opportunity of increasing their representation at the expense of aggressive individuals if no animals are reproducing more than others.

All the individuals in a captive population are descendants of the "founders"—those individuals originally captured in the wild. Founder animals are the source of all genetic variability in the captive population, and a population with many founders will be more genetically diverse. Yet the

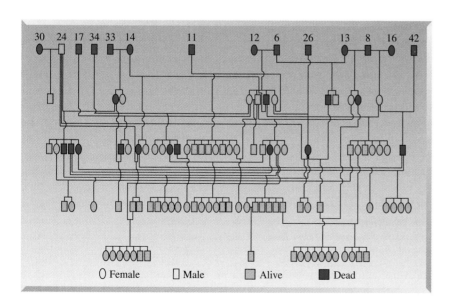

◄ This pedigree for the world's entire red wolf population on December 31, 1988, traces the ancestry of each red wolf then alive back to two of the 14 founding members of the captive population. The numbers at the top are the founder IDs.

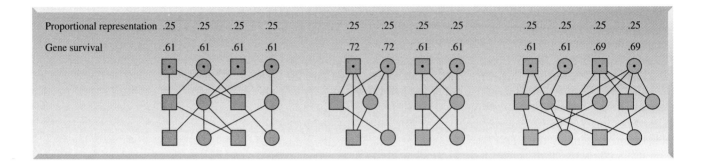

| Proportional representation | .25 | .25 | .25 | .25 | | .25 | .25 | .25 | .25 | | .25 | .25 | .25 | .25 |
| Gene survival | .61 | .61 | .61 | .61 | | .72 | .72 | .61 | .61 | | .61 | .61 | .69 | .69 |

▲ These three simple pedigrees illustrate the effects of inbreeding and family size. In all three pedigrees, the proportional genetic representation in the surviving population is the same for each founder (the founders are indicated by dots). But the probability that any one founder gene survives starts to vary when there is inbreeding (middle pedigree) or a difference in family sizes (right-hand pedigree). The danger is that genes will come to dominate in the captive population that did not dominate in the wild.

number of founders is often pitifully small. The 63 Californian condors currently in captivity are descended from only 14 founders, and the captive (and reintroduced) black-footed ferrets from 9. The Asiatic wildhorse (Prezwalski's horse) are the descendants of only 13 founders, as are the captive red wolves. When populations are this small, as much of the remaining genetic variability as possible must be conserved. Since each individual born in captivity obtains half its genes from each of its parents, it is very important that each parent produce more than one offspring. The more offspring an animal has, the likelier that genes not passed on to one offspring will be passed on to another.

A potential founder animal cannot contribute to genetic variability if it doesn't reproduce. If only one or two of the founders produce most of the offspring, then the population will not maintain variability no matter how many animals were originally brought in from the wild. Unfortunately, it is not rare for only a few of the captured animals to contribute most of the offspring. Consider the contributions of the 16 wild-caught Guam rails. Before being brought into captivity in 1984, this flightless bird from the island of Guam had been almost completely wiped out by a voracious snake that had invaded the island soon after World War II. Five of the 16 potential founders have not contributed any offspring at all, while three have made very high contributions, and within several years their offspring accounted for more than 40 percent of the captive population.

Genetic Management in the Absence of Pedigrees

Complete pedigrees are only available for a few captive populations and almost no wild populations. Although we may soon be able to develop more complete pedigrees using DNA fingerprinting, this type of analysis is expensive and only likely to be used in cases where a particularly valuable species has been reduced to very small numbers. Instead, genetic management of most populations must rely on information about the population as a whole rather than on data about individuals. The key statistic used in this type of management is effective population size, N_e, an estimate of the number of unique genetic individuals in the population contributing offspring to subsequent generations.

It is almost inevitable that a population of captive animals will lose some genetic variability, but this genetic variability will be lost increasingly more slowly as the effective population size gets larger. Moreover, a fast-growing captive population will lose a smaller total amount of genetic

Estimating Effective Population Size

The effective population size of any population will be determined by both its population size and its social system. In monogamous species, such as gibbons and parrots, everybody who survives to reproductive age gets to reproduce at approximately the same rate. For such species, the effective population size is given by a relatively simple formula:

$$N_e = \frac{4 N_m N_f}{N_m + N_f}$$

Here N_m is the number of breeding males in the population and N_f the number of breeding females. Thus in a population with 20 adult males and 20 adult females, the effective population size is 40, exactly equal to the actual population size. However, many species are polygamous, and in these species only a few males dominate the mating. This formula can still be used to examine, for example, the case in which all the females produce offspring, but only three of the males mate with the females. Under these circumstances, our estimate of N_e decreases to just over 10, less than the number of females!

A variety of our factors can further reduce N_e. In particular, variability in the number of offspring that each female produces will lower the effective population size. The ratio of effective population size to actual population size, N_e/N, is generally used as an index of genetic variability in a population. For the species for which data is available, the N_e/N ratios tend to fall in the range 0.2 to 0.5. As we would predict from the simple formula described above, populations of the more polygamous species have values at the lower end of this range.

variability. It is therefore best to increase the captive population to the maximum size possible in the circumstances as rapidly as possible. One potential obstacle, though, is that every animal may not be compatible with the individuals of the opposite sex it has found itself in captivity with. Furthermore, the zoo must provide the animals with suitable facilities for mating and rearing their young. The animals' needs may not always be obvious, as the experience with red pandas illustrates. Red panda cubs in captivity were dying at a very high rate until John Gittleman of the Smithsonian Institution realized that many of the cubs were being killed by neck injuries that took place as their mothers carried them around. When disturbed in the wild, red panda mothers move their young to a new den. When disturbed in captivity, and supplied with only one den, they carry their young around vainly searching for a new nest site. After Gittleman spotted the problem, zoos began providing their pandas with a number of alternative nest sites. Cub mortality declined and the captive red panda population began to expand.

Assuming we can find innovative ways to stimulate captive populations to increase at a rapid rate, we now need a rough estimate of the captive population size that we should aim for. The best size is the one that minimizes the rate at which genetic diversity is lost from the population if we can prevent selection from operating. The basic theory of population genetics predicts that genetic diversity is lost at a rate that is inversely related to the effective population size. Obviously it would be good to raise as large a population as possible, but zoos must consider their limited space and resources. The question for zoos is, what size population do they *need* to achieve to preserve a predetermined percentage of genetic diversity? That size turns out to depend to some extent on the physical size of the animal and on its social system.

To a good approximation, the percentage of genetic heterozygosity (an important measure of genetic variability) lost per generation roughly equals $\frac{1}{2}N_e$. Therefore, if the breeding program is to preserve at least 75 percent of the initial genetic diversity carried into captivity, it should last no more than $0.6N_e$ generations. If effective population size is 50, the population can be kept in captivity for 25 generations, whereas if N_e is 25, it should only be kept in captivity for 12 generations. There is thus an important difference between captive breeding programs for large species such as bison, which have a generation time of five to eight years, and those for smaller mammal species such as the striped grass mouse, which has a generation time of less than a year. A large population of mice will lose its genetic heterozygosity at a much faster rate than a smaller and slower-reproducing bison population. To maintain the striped grass mouse in captivity for two hundred years would require a population of 1275; to maintain bison in captivity for the same length of time would require a population of fewer than 130.

▲ Red panda in the Himalayan highlands exhibit at the Bronx Zoo.

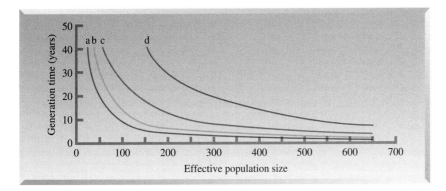

A larger effective population size is needed to maintain diversity when (1) the generation length is short and (2) the number of founders is small. This graph illustrates the effective population size needed to maintain 90 percent of the founder genetic diversity for two hundred years, when there is no founder effect (curve *a*), or when there are (*b*) 20 founders, (*c*) 8 founders, or (*d*) 6 founders.

The Problem of Hybrids

The presence of baby animals is always a major attraction to people who are thinking of visiting a zoo. One large city zoo expects its annual gate to increase by as much as a million dollars in a year when its polar bears produce cubs. Until quite recently the popularity of baby animals created an incentive for some zoos to produce charismatic babies from combinations of parents that might not strictly meet in the wild. The orangutan is an example of a species often subjected to such haphazard matings in the past. There are two subspecies of orangutan, one from Borneo and the other from Sumatra. Before the 1970s zoos simply mated orangutans without regard to their subspecies, so that there are now as many 80 hybrid orangutans on display. At present the SSP for orangutans will not allow the hybrids to reproduce; instead they will be maintained on display until they die and the hybrids become extinct.

Other hybrids have proved more controversial. The red wolf (*Canis rufus*), for example, is a natural hybrid between the grey wolf (*Canis lupus*) and the coyote (*Canis latrans*). It lived originally in the southeastern United States, where it became extinct in the wild earlier this century. An extensive captive breeding program has been established for the wolf, and the animal has been successfully reintroduced into the wild in a number of locations. Nevertheless, Bob Wayne and Sue Jenks of the Institute of Zoology at London Zoo suggest, from their review of DNA studies and morphological similarities in structure and appearance, that the red wolf is not significantly different from coyote populations living in areas from which the wolf had disappeared. This news greatly distressed zoos and federal agencies, which had spent considerable funds on the captive breeding programs for the species.

Wayne and Jenks's work raises an interesting additional problem. Many new plant and invertebrate species arise through hybridization—although many hybrids are sterile or die early in development as the result of

▶ A male of the Sumatran subspecies of orangutan (left) and a male of the Bornean subspecies (right). The Sumatran males typically have more whiskers and redder fur, while the Bornean males have larger cheek pads and darker fur.

incompatible combinations of genes, some are very successful and spread quickly. If red wolves are an early stage in the evolution of a new species, we may prevent their further evolution by deciding to abandon captive propagation programs for the animal.

Reintroduction and Translocation

The Yucatan peninsula in southern Mexico, Guatamala, and Belize is home to the black howler monkey *(Alouatta pigra)*. Deprived of its natural habitat by the destruction of Central America's rainforests and hunted for food, the black howler monkey has severely declined in numbers. The species is now classified as threatened by CITES and "insufficiently known" by IUCN. One of the last areas to maintain healthy populations of the species is a stretch of land along the Belize river. Recently, the establishment of a new reserve in the Cockscomb Basin Wildlife Sanctuary has provided an opportunity to develop a translocation program to bring howler monkeys into an area that contains a few of the monkeys already, but lots of suitable habitat for a larger population.

In 1991, Fred Koontz of the New York Zoological Society and colleagues from the Community Baboon Sanctuary in Belize began a project to take black howler monkeys from a healthy growing population found along the Belize river and move them into an area of the Cockscomb river basin that had previously been populated by howler monkeys but was now empty of them. Three release sites were selected, each close to trees known to be sources of food to howler monkeys. Individuals taken from the same troop would be released together at the same site. The sites were located one kilometer apart, sufficiently isolated to prevent stressful interactions between troops.

To obtain the animals to be relocated, 14 individuals from three troops of howler monkeys were darted with an anesthetic. Blood was taken for genetic analysis, and each monkey was photographed and marked with a distinct collar for later identification. The monkeys were moved overnight from the Belize river to the release sites in the Cockscomb river basin, where they were held for two to three days in chicken-wire cages to check that no animal was injured and to allow the troops to adjust to their new location. Once released, the troops were monitored every two or three days by locating individuals that had been fitted with radio collars. After 10 months of observation, 12 of the 14 released animals were known to be still alive and 4 infants had been born. Their numbers have since been supplemented by additional translocations. The project is an important success, but these animals had spent almost no time in captivity. How does this case compare with cases where animals have been kept in captivity for many years?

Once a captive population is producing significant numbers of new individuals, and if the forces that caused its decline have been removed or modified, it is sensible to begin thinking about reintroducing its members back into the wild. Mark Stanley-Price of the African Wildlife Foundation developed many of the principles underlying a successful reintroduction in his work on the Arabian oryx—a species of antelope that by the 1960s had been hunted to extinction in the deserts of the Middle East. This work suggested that a feasibility study is crucial to determine if the agents that reduced the population in the first place are still operating, whether suitable habitat remains, and whether funds are available to do the job properly. In the case of the oryx, the reintroduction project was judged feasible once guarantees had been provided that no hunting would take place in a large plateau at Yalooni in Oman.

Once this suitable site for liberation had been identified, a suitable group of animals for release had to be chosen and an appropriate strategy for this reintroduction developed. Eighteen oryx were selected for the initial introduction out of more than 400 captive animals; all had been bred in captivity in the United States and had lived in habitat at the San Diego Wild Animal Park that was similar to that at the release site. Determining which animals to reintroduce is not easy, since there are many different ways one can select 18 animals out of a population

▲ An Arabian oryx with its young, bred in captivity. The oryx are easy to find and hunt in the wild, because they do not move rapidly in the desert heat and they often stand conspicuously on hillocks to avoid being lost by the rest of the herd. When the oryx had been hunted almost to extinction, the last three surviving individuals were taken from the wild and, with another six individuals brought in from zoos around the world, formed the founder population of only nine individuals.

of 400! However, some animals effectively exclude themselves through their overfamiliarity with humans or overdependence on their captive diet.

The actual release took place gradually as the enclosure surrounding the animals, which started off as a holding pen, was slowly widened into an outside enclosure and then removed altogether. After spending two years in the enclosures, enough time to form a coherent social group and to become acclimated to the desert, the first herd was released on January 31, 1982. The release date was carefully chosen: concerned that the animals might break away from the herd and run, Stanley-Price and his colleagues chose the winter date to avoid the danger of heat stress and because the poorer grazing in that season ensured that the herd stayed together and close to the compound to get their food supplements.

The local Harasi tribesmen were encouraged to play a highly active role in the program, as protectors, wardens, and observers. They have benefited from the presence of the oryx in both a material sense (because they are paid wages for their stewardship of the animals) and in a spiritual sense. When the oryx first returned to Yalooni, the sceptical elder tribesmen were delighted to see the same animals they remembered from their youth, and the younger men were amazed at this materialization of a mystical element in their cultural history. The reintroduced population was monitored almost every day for several years so that information could be gleaned for future reintroductions. Four years after the initial release, one herd had a range of 2000 square miles, within which were a number of favored grazing sites separated by distances of up to 40 kilometers. A second herd was subsequently introduced at Yalooni, and other herds have been established in Quatar, Abu Dhabi, Jordan, Saudi Arabia, Bahrain, and Dubai.

When considering the overall success of reintroductions, we need to differentiate between translocation, as exemplified by the howler monkey case, and reintroduction following captive breeding, as exemplified by the oryx. Translocation is the release of animals taken from the wild directly back into the wild at another location in an attempt to reestablish or augment natural populations. Brad Griffiths and his colleagues at the University of Idaho have surveyed the nearly 700 translocations and reintroductions of native birds and mammals conducted in Australia, Canada, Hawaii, New Zealand, and the continental United States between 1973 and 1986. This survey suggests which features of species biology or translocation habitat favor success and which court failure. Native game species are likely to be successfully translocated; threatened or endangered species are not. Moreover, the success of the translocation is very sensitive to habitat quality. When a species is translocated into the core of its historical range, success is more likely than when it is translocated to the periphery of or outside the historical range. Similarly, herbivores seem to do better when translocated than carnivores or omnivores. Successful

Reintroducing the California Condor

The first two Californian condors were reintroduced back to the wild in the summer of 1992. They were followed by four birds in the late summer of 1992 and several other birds since then. One of these birds died when it struck a power line, two others that failed to adjust to the wild had to be recaptured, but the others are still out in the wild. Released in the Los Padres National Forest, the birds range 150 miles from Santa Barbara to San Bernadino. Although the birds sometimes manage to find their own food, they mainly live off food provided by researchers, who don't want the birds feeding off lead- and pesticide-contaminated carcasses.

The reintroduction was carefully prepared for. First the terrain was tested: before the condors were released, 13 captive-hatched Andean condors were let go in Los Padres National Forest in three separate releases beginning in December 1989. Only female Andean condors were freed to prevent them from becoming established in North America. They seemed to do well, staying in the safe mountainous areas and avoiding contaminated food in favor of meat supplied by researchers.

The condors that were released were chosen from the 52 that were in captivity. The main criterion for their selection was that they be individuals who had produced sufficient offspring that they were already well represented in the gene pool of captive birds. The selected condors were released in the company of four Andean condors, as the gregarious Californian condors are more likely to thrive if they have company. Six other Californian condors, all hatched in 1992, were released into the wild that December. The major serious concern of a genetic nature facing the condors is that all 63 Californian condors are descended from 14 original founders. These can be divided into three distinct groups or clans based on their genetic lineage. Birds within each clan are highly inbred but they are less so across the clans.

By the spring of 1993 there were 63 Californian condors alive, 56 at the San Diego Wild Animal Park and Los Angeles Zoo, and 7 in the wild in Los Padres. There are now plans to release more birds into part of the Grand Canyon.

translocations release more animals than unsuccessful translocations, and where the animals being introduced have been captured from the wild, success is higher if the animals have been taken from a dense and growing population rather than from a population of medium or low density or stable, even declining numbers. Overall only 44 percent of translocations of threatened, endangered or sensitive species were successful, with wild-caught animals tending to be considerably more successful (75 percent) than translocations with captive bred animals (38 percent).

It seems always more sensible to carry out multiple reintroductions, at several sites, rather than to rely on a single reintroduction. Yet even a successful single reintroduction can require quite large numbers (more than 100 individuals) to be released over a period of time. Because reintroducing small numbers of endangered or threatened species, even in excellent habitat, is a shaky proposition, it is clear that translocation must be considered long before it becomes a last resort for these species, essentially before density becomes low and the populations are in decline.

Captive Propagation of Plants

In the United States at least, far more attention is given to the captive breeding of endangered animals than to the breeding and reintroduction of the approximately 5000 species, subspecies, and varieties of equally endangered plants. The number of plant species threatened with extinction is particularly staggering because it represents about 20 percent of the native flora of the continental United States, Hawaii, and Puerto Rico. Many of these plants are naturally confined to small areas that are being lost to construction, agriculture, and housing projects. Plant species are often finely adapted to the soil and climate of their habitat; thus, even more so than animals, plants are closely tied to the local conditions in which they grow. Ultimately their conservation must be based on preserving the land to which they are adapted.

The Center for Plant Conservation (CPC) was created in 1985 to encourage scientifically credible plant conservation at botanic gardens and arboreta in the United States. The original network of 18 gardens was selected to represent as many geographic regions of the country as possible, so that just about any plant species could be grown outdoors if necessary. By 1990, five years after CPC was founded, the member gardens, now up to 20 in number, maintained collections of over 300 of the nation's rarest plant taxa, and the number continues to grow by 50 to 100 taxa each year.

Ironically, plants may be better suited to captive propagation than the animal species that have formed the main focus of captive breeding programs. In many cases, we don't have to keep whole living plants, but simply their germ plasm—the reproductive cells in pollen or seeds. The germ plasm of some plants may be conserved simply by storing the seeds in refrigerated "seed banks." The method works best with small seeds of low oil content from temperate regions, where plants are adapted to periods of winter dormancy. When properly stored at subzero temperatures, the seeds may remain viable for many times longer than they would in the wild.

Seed banks are perhaps the only efficient method of preserving annual and short-lived perennial species for a long period. Because their lives are short, these plants must be bred frequently, and frequent breedings would make genetically static populations logistically impossible to maintain. In contrast, long-lived herbaceous perennials, shrubs, and trees are better stored as living collections. Where space is no object, a sufficient number of plants can be kept for a long period, in some cases up to several hundred years, while genetic diversity is maintained by avoiding sexual reproduction, through which rare alleles may be lost, in favor of cuttings and other nonsexual propagation methods.

When the time comes to reestablish plant populations in the wild, it is insufficient for plants or their seeds to be simply placed in the soil and forgotten. As with animals, newly introduced plants need constant monitoring; often a number of attempts to introduce the species are necessary before one "takes." In some cases plants have to be introduced to a number of different localities, particularly if climate change has moved the region where the temperature is optimum for the species.

The Economics of Captive Breeding

Although the success of reintroductions, uneven at present, will improve with experience, the concept of zoos as arks suffers from a number of other serious shortcomings. The costs are significant, and the space limited. William Conway, Director of the New York Zoological Society, has calculated that perhaps a maximum of one million individuals can be maintained in the world's approximately one thousand zoos. If healthy populations of around 250 to 500 individuals of each species are kept in captivity, then only around 1000 to 2000 species could be saved this way.

The costs of captive propagation are related to the social system of the species. Group-living species such as the Arabian oryx are relatively cheap to keep in captivity; in the main they require only a paddock with suitable pasture and food supplements. In contrast, even small territorial species such as the striped grass mouse are much more expensive, since each individual requires an enclosure that provides shelter from adverse climate, predators, and aggressive members of its own species! The problem is actually compounded for the smaller territorial species, since, with their shorter generation times, they need to maintain a larger effective population size during their time in captivity. Whereas a zoo or conservation center can readily promote its herd of Arabian oryx as an attractive spectacle for visitors, it is unlikely to set aside several buildings to hold the cages of identical exhibits, each containing the same species of mouse.

In many cases the economic costs of captive breeding far exceed the costs of conservation in the wild. For example, Nigel Leader-Williams has estimated that the annual cost of protecting rhinos in the wild (estimated at $575 per individual or $230 per square kilometer for black rhinos in Africa in 1980) is around one-third the cost of maintaining them in a zoo. Nevertheless, over the past ten years, a partnership of zoos has apparently spent $2.5 million trying to save the so-called "doomed" Sumatran rhinos, with

Comparative Annual Upkeep of 11 Animal Populations

Species	Generation Time (years)	N_e [a]	Animal Upkeep (U.S. $ 0,000)
Siberian tiger	7	136	57.7
Indian rhinoceros	18	53	11.4
Nyala	8	115	15.3
Striped grass mouse	0.75	1275	13.1
Brush-tailed bettong	6	159	9.0
Mauritius pink pigeon	10	95	8.7
Arabian oryx	10	95	7.8
African black-necked cobra	10	95	6.3
Bullfrog	7	136	4.0
White-naped crane	26	37	3.1
Caribbean flamingo	26	37	1.1

[a] The populations were maintained at an effective population (N_e) sufficient to sustain 90 percent of founder heterozygosity over 200 years.

◀ The Sumatran rhinoceros is one of five rhino species, each very different from the others. Threatened by poaching and deforestation, its population in the wild, standing at several hundred individuals, has become fragmented into small patches of remaining habitat.

the aim of establishing captive populations of this species in zoos. This sum does not include the cost of maintaining the animals once captive. Losses of rhinos during capture (three deaths) and particularly after capture (six deaths) have been high. By May 1991, 21 rhinos were in captivity, and there have been no births, except to one female who was pregnant when captured.

In contrast, if it costs around $230 per square kilometer each year to protect rhinos in the wild, then $2.5 million could protect 700 square kilometers of prime rhino habitat for nearly two decades. An area of that size would hold around 70 Sumatran rhinos (at a natural density of one rhino for each 10 square kilometers). If the rhino population grew at a rate of around 0.06 percent per year, the 70 animals in the reserve would give rise to 90 calves in this 20-year period. The Minnesota zoo defends its captive rhino program by pointing out that the zoo helps support Ujung Kulon, the 300-square-mile national park on Java's eastern tip that holds the world's remaining Java rhinos. Although the $25,000 denoted to Ujung Kulon annually is only a small fraction of the zoo's annual receipts, it provides about a third of the park's operating budget.

Conclusions

Most of the economic processes that are damaging natural habitats are encroaching on wild populations and communities much faster than these populations can adapt to the consequent changes in their environment.

Taking at least a part of the population into captivity may be the only alternative to extinction, but it brings with it its own formidable set of problems. When species are removed from their natural habitats, natural selection ends and selection for the captive environment begins. When wild animals are denied the opportunity to choose and compete for mates, genes from males that might not obtain an opportunity to breed in the wild will be represented in subsequent generations, risking a loss of fitness.

We are learning much about how to keep endangered species in captivity, but there is still much we need to know, and many problems to be resolved by ecologists, geneticists, and the curatorial staffs of zoos and other conservation centers acting together. Chief among their tasks is to acquire additional scientific knowledge about how levels of genetic diversity change as the size of a captive population rises and falls. This information is important for the management of not only populations in captivity, but also the small populations of free-living species in wildlife reserves and elsewhere. As the natural habitats of free-living species become increasingly fragmented, their populations, too, will become split into smaller groups, each potentially faced with the same problems of inbreeding and missing genetic diversity that trouble zoo populations. We need to know how the persistence of increasingly fragmented populations will be affected: Could, for example, there be a sufficient gene flow between these populations, through fortuitous matings, to maintain genetic diversity?

A species cannot be reintroduced into the wild unless natural habitat exists to receive it. Plans for captive propagation and release will ultimately only be successful if they proceed in conjunction with projects that maintain the integrity of the natural habitat in the wild. We should not mistakenly believe that a population can survive forever in captivity. To regard captive maintenance of a species as conservation is to commit the conceptual error of assuming that species are somehow separate from their habitats. That sort of thinking seems to be at work in the case of the Hilo top minnow of Hawaii, for example. Since artificially maintained habitats have been stocked with captive-bred fish, the species has been reclassified from endangered to threatened. Yet the species' main habitat in the wild continues to deteriorate, and all the other species that exist in this habitat have lost the benefits provided by the legal protection of a potential umbrella species.

When rapidly growing captive populations begin to exceed a zoo's capacity but habitats in the wild provide no opportunity for reintroduction, one answer may be the development of zoos and conservation centers in the species' country of origin. Luckily some of the world's leading zoos and botanical gardens have recognized this solution to the problem of overcrowding. The Duke University Primate Center has been exemplary in this regard: in addition to its original collection of lemurs located near campus in North Carolina, the center has developed a parallel collection of the animals at the Park Tzimbazas in Antanananarivo, the capital of Madagascar. A visit to the park provides the sole opportunity for many people in Madagascar to see the lemurs that are unique to their island and a highlight of its biological diversity.

As similar initiatives are begun to set up zoos in Nairobi and Belize, zoos, botanical gardens, and aquaria seem poised to play a major role in conservation education. Curators are constantly experimenting with novel ways of presenting a strong conservation message to zoo visitors. Although not everyone reads the graphics in zoos, the labels that zoo curators write describing their animals are still read by millions of people annually. (In contrast, very few scientific papers are read by more than 10 people.) Each year more than 600 million people visit the world's 1200 zoos and aquaria. With this formidable constituency, zoological societies seem to be more logical organizers of local conservation programs than universities or schools.

Unfortunately, more lions and cheetahs are seen by American children in zoos than are seen by African children in the wild. Certainly, many more children living in the Bronx have seen a snow leopard than children in Nepal. In 1990, U.S. zoos and aquaria sponsored 388 education, conservation, and research projects in 63 countries worldwide, a fifth of these involving the reintroduction, rehabilitation, or rescue of endangered species. More of these international programs could set out to develop zoos in countries where they could give children and adults a better understanding of the wealth of nature that surrounds them. There is something deeply sad about visiting remote villages in Madagascar or Brazil and finding the village children watching Jurassic Park or Rambo in their local video hut, while the forests around them are destroyed. They are swallowing a condescending myth, while their natural inheritance is sold for a pittance.

Short-term successes in captive propagation provide false reassurance that the endangered species problem is solvable through technological means. Many experts hope that in the future most captive breeding schemes will assume sustainability of the captive population in the wild as their ultimate goal. Unfortunately, many human beings continue to be captivated by the belief that we can easily replicate nature in captivity. The Biosphere-2 "experiment" in the Arizona desert is a bizarre monument to this naiveité: the biosphere requires constant maintenance, and the artificial communities designed to support life in the structure often become unbalanced. Although simplified versions of Biosphere-2, such as the "Ecotron" facility at Imperial College in London, will be invaluable in allowing us to discover how simple captive ecosystems can be managed, all the preliminary results from this work continue to illustrate that the best way to conserve species and natural ecosystems is in the wild.

Identifying Land for Nature Reserves

It's possible to leave New York as the sun is rising, fly west for four hours, change planes, and arrive in Bozeman, Montana, in time for lunch. You can then drive southeast until you reach the Yellowstone river as it runs down Paradise valley, in the shadows of the Absaroka and Beartooth mountains. Follow the river's course back up the valley and on through Yankee-Jim canyon and you arrive at the gateway to Yellowstone National Park with plenty of time to spare for a late afternoon hike. In less than a day you've made a journey it took early explorers many months to achieve. The pioneers who explored the area in the 1870s were interested in furs and possible mineral deposits, but the natural beauty and isolation of the area encouraged them to designate Yellowstone as the first national park to be established in the United States.

The major motivation for establishing Yellowstone as a national park was the unique geology of the area; its famous hot springs and geysers represent more than 50 percent of the world's active geothermal features situated on land. Most of the wildlife

◀ The northern entrance to Yellowstone National Park in Gardner, Montana.

163

species found in Yellowstone today—bears, moose, and elk—could still be found as far west as Pennsylvania at the time Yellowstone was established. The huge agricultural development that has taken place over the last hundred years has left Yellowstone and other national parks stranded as islands of wilderness in a sea of development and agriculture. They now hold the last remnants of the natural fauna and flora of the United States.

The first national parks in Europe were set up as early as the sixteenth century—England's New Forest was originally set aside as a game reserve by Henry VIII. Protected areas have existed in India since the fourth century B.C. Natural parks now exist in most of the world's countries; these range from the vast expanses of the Serengeti National Park in Tanzania to barrier reefs off the northeast coast of Australia and Belize to smaller reserves such as Berenty in Madagascar and Kimodo Island in Indonesia.

The character of a nature reserve is defined by the species and ecological processes within its boundaries, and these are essentially the same species and processes that define the underlying character of a country or region. By establishing nature reserves and wilderness areas, the people of a country maintain a set of natural values that are perhaps more fundamental in defining the country's identity than its political constitution and the system of moral or religious codes that determine its people's behavior. Unfortunately, it is only in the last twenty to thirty years that most countries have set aside land to protect their biodiversity, a considerably shorter time than the natural time scales of decades, centuries, and millennia on which ecological and evolutionary processes operate. In earlier chapters we described how land conversion is rapidly eroding the remaining natural areas of the world, particularly tropical forests. We have seen that we have little hope of conserving more than a few species in captivity. If we wish to maintain biodiversity in a natural state, it is essential that land be set aside in nature reserves. In this chapter we consider ways to determine which areas should be used as nature reserves and national parks.

Ecologists rely on a variety of approaches in designating land to be set aside as nature reserves, but most of these approaches assume that the principal goal of nature reserves is to preserve as great a variety of biodiversity as possible. Many people might argue that leaving an area untouched in order to preserve biodiversity is not the best economic use of land that might otherwise be mined, farmed, or logged for timber. Very little annual economic return can be expected from a large area of true wilderness that is not being exploited for its timber or mineral resources. However, large wilderness areas and natural forests perform a variety of ecological functions, such as the recycling and storing of water and carbon dioxide, that are essential to many human activities. The economic "value" of these activities is not easily quantified on an annual budget sheet. Nevertheless, small reserves with elaborate visitor facilities may provide their owners a substantial annual revenue and create a significant number of jobs in the local community. Yet small reserves that are managed for the tourist trade will only preserve a portion of the biodiversity in a region. Questions that relate to the management of nature reserves and their economic value are sufficiently important that I will return to treat them in detail in the following three chapters.

In approaching the question of designing nature reserves, I make the idealized assumption that ecologists will play a major part in deciding their size and location, that the discipline of ecology and conservation biology will provide insight to the politicians and land managers who decide where to place nature reserves. The actual state of affairs is quite opposite, and curiously paradoxical: although questions about the optimal size, shape, and number of nature reserves have greatly shaped the development of conservation biology as a science, very few of the world's major nature reserves were set up using insights that have emerged from this scientific discussion. Instead, governments have set aside land for nature reserves either in areas advocated by pressure groups lobbying for specific endangered species and communities or in areas subjected to minimal pressure from land developers or the expanding human population. Conservation biologists have developed their understanding of what constitutes a successful reserve system by performing retrospective, "post hoc" analyses, which attempt to elucidate the characteristics of those established nature reserve systems that have been most successful in conserving biodiversity.

Do National Parks Preserve Species?

The national parks in the western United States preserve many species that once had a much broader range in North America. Although the parks appear on the map as isolated islands, they are surrounded by public lands, national forests, and wilderness areas that in some cases extend the effective area of a park. In other cases logging and mineral extraction are disrupting habitats right at the park edge. How successful are these parks at preserving the wildlife within their borders?

An interesting and provocative analysis of the mammal diversity in the national parks of the western United States was undertaken in the mid-1980s by William Newmark while a graduate student at the University of Michigan. Newmark showed that there was a fairly good relationship between the area of a reserve and the number of mammal species found within it. The theory of island biogeography described in Chapter 2 conjectures that the number of species present in an area is determined by a balance of immigration and extinction. Because the National Park Service keeps detailed records of the species seen in the parks, Newmark was able to examine the colonization and extinction rates in a number of parks, of different sizes and degree of isolation. His analysis showed that there had been an excess of extinctions since the parks were established, and that the extinction rate depended on the size of the park, with extinctions increasing in parks smaller than 20,000 square kilometers.

Newmark's results were controversial when first published, partly because the National Park Service felt it was being unduly criticized, partly because Newmark had to restrict his analysis to the subset of mammal species for which data were available. Nevertheless, further analysis has suggested that most of the trends he observed are real, and mainly reflect the increasing extinction rates in parks smaller than 20,000 square kilometers. The decline in species numbers is further exasperated by the falling rates of recolonization: parks are becoming increasingly isolated, and further removed from any wildlife that could move in from the outside, as the

▶ The 20 national parks shown as colored areas are scattered like islands across the western United States. In all but the largest of these parks, more species are disappearing than are moving in from the outside.

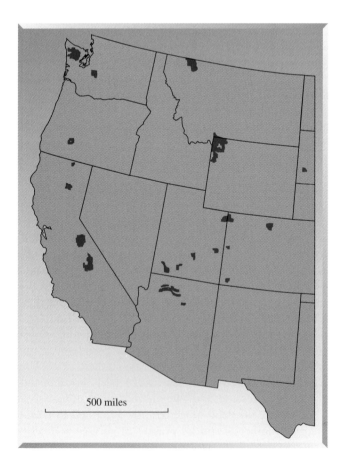

500 miles

areas around them are either clear-cut, overgrazed or colonized by towns and agriculture. The same trends have been found to characterize national parks in East Africa—mammal species are going extinct at higher rates in parks that are smaller than 20,000 square kilometers. Parks in the United States and East Africa may not be adequate in number and size to preserve their nations' wildlife; is that also true in the rest of the world?

The land set aside to protect biodiversity in other areas of the world seems to have increased by a factor of 8 since 1940—an increase from an area the size of Madagascar to an area larger than the Indian subcontinent. This area, comprising more than 8000 protected areas, covers 7.7 million square kilometers—approximately 5 percent of the earth's land surface.

At first sight conservation biologists would appear to have done an excellent job: the total area of land set aside to protect biodiversity since 1940 far exceeds that exchanged in all military conflicts during the same period, and its protected status has been accomplished at a minute fraction of the cost (less than 0.01 percent). However, the land set aside to protect biodiversity is only a small fraction of the total area of natural habitat that is being converted to agriculture or harvested for timber. The urgency of the problem becomes apparent if we examine the net amount of habitat lost and habitat preserved in parks and wilderness areas in key regions of tropical Africa and the part of Asia containing Indonesia and Malaysia, indicated in the figure on the next page. Notice that it is not only tropical forests that are disappearing, but also savannas and wetlands. Almost no savanna is protected in Indonesia and Malaysia, and very little wetland habitat in Africa. Although 1 to 2 percent of the original forest has been designated as protected in both these tropical realms, this total area is only slightly greater than the area of forest that is lost each year.

Furthermore, the patches of land set aside to protect biodiversity are disconcertingly skewed in size: small areas greatly outnumber larger areas, and the commonest size for a protected area is only around 10 to 30 square kilometers. Only 3.5 percent of parks are larger than the minimum size predicted by Newmark to have colonization rates that exceed extinction rates. Nevertheless, the majority of the world's protected land is concentrated in these few megaparks. Unfortunately, some of the largest parks are found in vast expanses of desert or mountain, which contain unique collections of species but at very low densities. Because deserts and mountains are unsuitable for agricultural purposes, it is easy for governments to declare regions within them as nature reserves.

Obviously, it is dangerous to extrapolate from studies of parks in North America and East Africa when drawing conclusions about parks throughout the world. Colonization and extinction rates will certainly be very different in tropical parks on different continents; so perhaps an optimistic 5 to 10 percent of national parks may be large enough to maintain their present level of biodiversity. Nevertheless, it would be dangerously

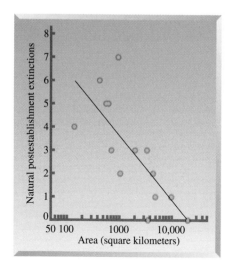

▲ William Newmark found that, for the 14 parks described by this graph, more species had gone extinct in the smaller parks since their establishment than in the larger ones. The number of extinctions decreases fairly steadily with the size of the park.

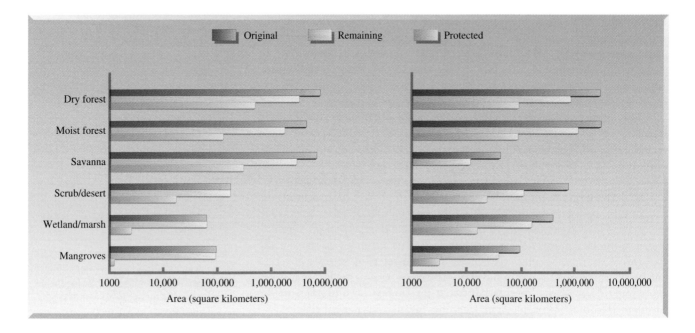

▲ Forests and savannas in particular are losing area in tropical Africa (left), whereas all types of habitat have experienced considerable loss of area in Malaysia and Indonesia (right). Don't be misled by the similarity in lengths of the three bars, which is a result of the logarithmic scale used for area. For example, only a quarter of tropical Africa's original dry forest is still standing, and of this remaining forest only a tenth is protected.

naive to assume that we already have enough parks; to do so would be to ignore the extinction of species not yet in national parks, while taking for granted that all national parks offer comprehensive protection to all the species they contain.

This initial overview suggests that certain types of habitat are underrepresented in some reserve systems and that many protected areas may be too small to maintain healthy communities of plants and animals. Are there any ecological methods that we could bring to bear to identify remaining land areas that could be set aside to protect the representative flora and fauna of a region's various habitats?

The SLOSS Debate

The success of MacArthur and Wilson's theory of island biogeography suggested to the evolutionary biologist Jared Diamond, of UCLA Medical School, that the theory could be applied to nature reserves. Diamond had observed that the pristine areas of habitat still available to wildlife were increasingly becoming "islands" in a landscape dominated by urban and agricultural activity. This insight prompted Diamond in 1975 to propose a set of guidelines, suggested by MacArthur and Wilson's theory, for the shape and position of nature reserves. The general spirit of these guidelines was to suggest that the size, shape, and degree of isolation of a patch of habitat would all be important in determining how well it served as a refuge for communities of threatened or endangered species. Thus large reserves would conserve more species than small reserves, and reserves that are situated close together, or are linked by corridors, would tend to hold more species than reserves that are isolated far from others.

Unfortunately, Diamond's insightful caricatures were occasionally interpreted as literal designs for nature reserve systems. More significantly, his guidelines were criticized on the grounds that a system of reserves that maximizes the number of species in the system might not guarantee the survival of any particular endangered species. These differing points of view evolved into the SLOSS debate, so called because it addressed the question whether a *single large* or *several small* reserves maximized diversity within a nature reserve system.

If we consider only the basic details of the species-area curves, then it seems simplest to suggest that procuring the single largest area that we can afford is the best strategy for maximizing species diversity. However, this simplest case ignores a basic fact: it is unlikely that any single reserve will contain the full diversity of habitats available to us within even a small country. More disconcertingly, if expanded to a global scale, then in its purest form this logic says "preserve only the largest continent!" Essentially, this approach ignores the considerable scatter of points that occurs around the underlying species-area relationship: simply making a reserve larger won't increase the number of species in *every* case. A smaller site that encompasses a greater diversity of habitats could contain more species than a larger, but more uniform site. If we want to design a reserve system to maximize the number of species we save, we need to extend our approach to consider geographical variations in the types of species and habitats.

An alternative strategy, which maximizes the diversity of habitats, is to buy several smaller patches of land to be set aside as nature reserves. Suppose a conservation organization has a fixed sum of money to spend on land acquisition. If we assume that the price of land *per unit area* is independent of the size of the patch of habitat, should we spend our money on one large nature reserve, or will we be able to protect a larger number of species by buying a variety of smaller pieces that together have the same net area?

Whether we think a system of small reserves is better than a single large reserve depends on whether we consider the consequences of subdivision for a single species or for whole communities of organisms. One reason the SLOSS debate has been so difficult to resolve is that it has assumed that the answer to one of these questions also applies to the other—that a system that is best for one species is best for all species or, conversely, that a system that is best for most species is best for a particular species. Neither supposition is necessarily the case. A system of small reserves may provide a greater variety of habitats, but at a cost: although some species could

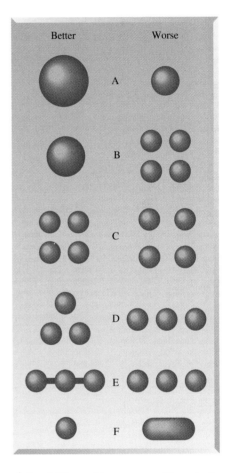

▲ Jared Diamond has compared hypothetical configurations for nature reserves, relying for guidance on the theory of island biogeography. Reserve designs labeled "better" should have lower rates of species extinction that reserves designs labeled "worse."

persist as metapopulations in a mosaic of different habitat types, in general habitat fragmentation is detrimental to the persistence of species, in particular those that require large areas of contiguous habitat.

To determine which strategy is the better of the two in any particular case we need to know something about the similarity of the species present in different habitat patches. Do different habitats contain almost identical collections of the same species? If so, then it won't be as important to preserve all types of habitat. Moreover, we need to know something about the slope of the species-area curve. When similarity between the communities is high and the slope of the species-area curve is steep, then it will be sensible to invest in one large nature reserve. In contrast, when the similarity between communities in different patches of habitat is low (as would be found in a more varied landscape) and the slope of the species-area curve is shallow, then the better choice will be to procure a number of reserves of intermediate size. In general, both the similarity of species composition and the slope of the species-area curve will vary with the geographical scale at which we examine the problem. On the local scale of a town or district, species-area curves will rise sharply, while many of the same species will be present from patch to patch. On the continental scale, species-area curves will have shallow slopes, while patches may be very different in species composition.

The optimal strategy evidently depends on whether we are a small local organization concerned with preserving biodiversity within our own limited area or a nationwide organization, perhaps a national government. If we are a small local conservation organization, we might better concentrate our resources and focus on saving one large piece of land. If we are a national or even international organization, we would instead diversify and procure a variety of nature reserves of intermediate size.

How Do We Identify Land for Nature Reserves?

In its simplest sense the design of nature reserves can be divided into two components: we have already discussed the problems of size and shape; we now need to consider location. How can we identify areas that are rich in biodiversity? This task is made difficult by our lack of information about the abundance and distribution of most animal and plant groups. In the absence of detailed information for a wide range of species, is it possible to use commonly studied taxonomic groups such as birds as surrogates for less well studied groups?

Perhaps the first study to truly tackle the problem was undertaken by John Terborgh and Blair Winter of Princeton University, who suggested using the diversity of endemic bird species as a means of identifying areas that should be set aside for national parks in Columbia and Ecuador. They identified a number of areas in the South American continent where endemic bird species were especially numerous. Studies of these areas revealed plant species to be unusually diverse as well. These areas of South America have probably been exceptionally rich in species since the Pleistocene epoch more than 10,000 years ago, when global changes in the climate caused large amounts of what is now the Amazon basin to dry out. Forests shrank until large patches of forest were restricted to only 20 to 30 areas of the continent. The "Pleistocene refugia hypothesis" posits that these forest patches were isolated for a sufficiently long period that the plant and animal species living in them evolved to produce a variety of species distinct to each patch. Once the climate became wetter, the patches could expand, and they eventually merged to form the present vast Amazonian forest. Although similar patterns have been found for butterfly, plant, and bird distributions in Africa, the Pleistocene refugia hypothesis is

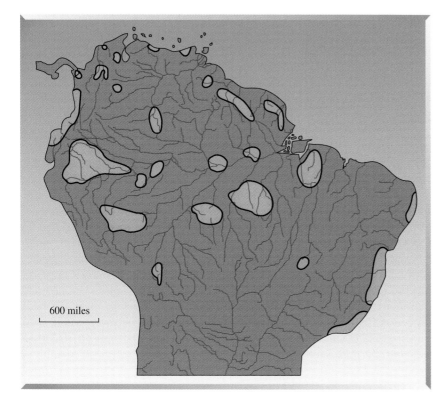

◀ The 26 colored areas are regions of tropical South America believed to have exceptionally high numbers of endemic species.

600 miles

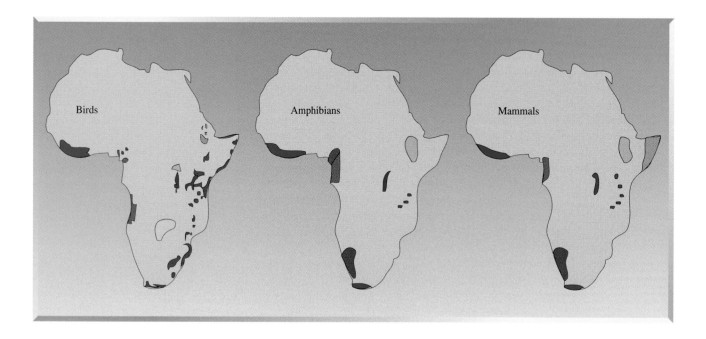

Birds   Amphibians   Mammals

▲ Centers for endemism for birds in Africa correspond in many cases to centers of endemism for amphibians and mammals. The correspondence suggests that areas that are high in diversity for one species may be high in diversity for others.

now considered to be only a partial explanation of South America's high species diversity, as alternative explanations have been found for some of the distribution patterns. In particular, many of the areas of high species diversity coincide with distinctive underlying soil and mineral depositions.

The strategy outlined in Terborgh and Winter's 1982 paper has since been adapted to accomodate alternative ways of identifying areas that are naturally high in diversity. In particular, the ornithologists at the International Council for Bird Preservation (ICBP) in England have used a computer data base of bird species distributions to identify Endemic Bird Areas (EBAs) throughout the world. Their analysis suggests that many of the world's rarer bird species are comparatively restricted in their distribution and that certain areas are particularly rich in endemic bird species. For example, 20 percent of the world's bird species with restricted ranges are confined to only 2 percent of the land surface. As the distribution of bird species tends to be more completely documented than that of any other group, the scientists at ICBP have suggested that planners target land to be set aside for nature reserves so as to maximize the number of endemic bird species. The hope is that they will thereby maximize the land set aside for other taxonomic groups as well.

This approach obviously assumes that where levels of diversity are high for one taxonomic group, they will be high for others. The evidence in support of this contention is ambiguous. On the one hand, hot spots of tiger beetle diversity, recorded over the entire continents of North America and Australia, seem to coincide with other types of biodiversity. On the other hand, there is little correlation between bird diversity and the diversity for other phyla in the United Kingdom. The differences between the results for tiger beetle diversity on a continental scale and bird diversity in the United Kingdom may reflect differences in geographic scale, but they may equally reflect differences in the quantity and quality of information available. The paucity of information is a second and perhaps more pressing problem for those who would like to use surrogate taxonomic data to identify crucial areas for nature reserves: in many parts of the world we still know very little about the distribution of species living in any area. As was suggested in Chapter 1, for many groups of organisms much of the basic taxonomic work remains to be done. To help address this problem, Conservation International has set up the Rapid Assessment Program, designed to take a group of experts into an area where they quickly assess the diversity of various taxonomic groups in a short period of time. Although initially very productive, the effort was tragically disrupted when, in the middle of one survey, a plane crash killed Alwin Gentry and Ted Parker, the team's botanical and ornithological experts.

▼ There is often a correspondence between tiger beetle species diversity and butterfly species diversity. The graphs plot the number of tiger beetle and butterfly species per square on grids across North America and Australia.

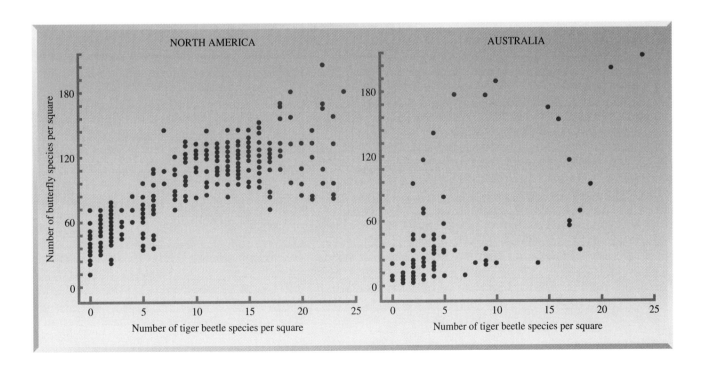

Gap Analysis

Computer-savvy ecologists have recently created a powerful tool for selecting areas of land that can be set aside as nature reserves. Called Gap Analysis, the technique is a step forward from the opportunistic approaches adopted in the past. It was inspired by the rapid development in the last ten years of computer-based geographical information systems (GIS), which produce maps showing features of the landscape ranging from topography to land ownership. Ecologists realized that some of the features mapped by GIS, such as land use patterns, were features that they needed to know about in planning the sites of nature reserves. If they could bring their own information about species diversity to the analysis, they would be able to search for gaps in patterns of land use where the levels of biodiversity were unusually high or the numbers of unique species unusually large. In this way, a Gap Analysis would seek to create a reserve system that avoids uneven representation of different species and habitat types. To achieve its goal, it concentrates not on finding single sites, but on setting aside combinations of sites with complementary features.

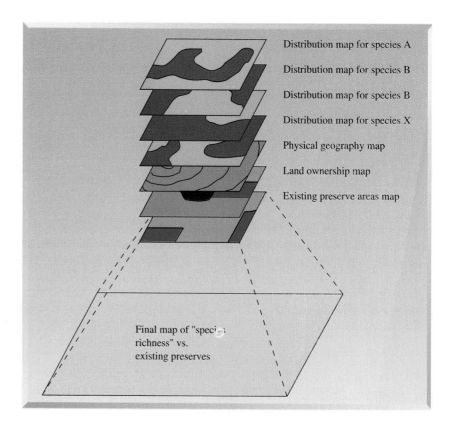

▶ A Gap Analysis superimposes maps of species distributions over maps produced by geographical information systems, to reveal areas of species richness not already included in nature reserves.

Distribution map for species A

Distribution map for species B

Distribution map for species B

Distribution map for species X

Physical geography map

Land ownership map

Existing preserve areas map

Final map of "species richness" vs. existing preserves

◀ A fence marks the boundary of Hakalau Forest National Wildlife Reserve, established high on the windward slope of Mauna Kea to preserve the area's 36 species of bird, including 7 endangered native species. The fence keeps out feral cattle and pigs, both the descendants of animals introduced by Europeans, and alien plants. To the left of the fence is land that has been overgrazed by cattle; to the right, in the reserve, cattle have not grazed in five years, and the grasses and ferns are growing more luxuriously. Eventually the park managers hope to reestablish the slope's original forest.

The first step in a Gap Analysis is to produce computerized GIS maps of the topography, vegetation, hydrology, and land ownership of a region. The next step is to generate computer maps showing the geographical distribution of the region's animal and plant species. These distribution maps are then superimposed onto the land use maps. The goal is to find areas where unusually high levels of species diverisity coincide with minimal human use of the land. The analysis also takes into account the species already safeguarded by existing reserves, so that new reserves can focus on species not yet under protection.

Although the Gap Analysis approach tries to identify a more diverse collection of species and habitat types, its results may still be biased if the analysis relies on information from only one group of species—sites that are superb for bird diversity may not necessarily be rich in plant or insect species. In many cases taxonomic groups that are not explicitly considered when setting up a reserve system will be protected only incidentally, if at all. For this reason, the approach is most effective in countries where amateur and professional naturalists have already gathered a wealth of natural history data.

The approach was first successfully developed in the mid-1980s in Hawaii by Michael Scott of the University of Idaho, who used it to compare the geographic distribution of endangered forest bird species with the distribution of existing nature reserves. The analysis instantly uncovered a flaw in Hawaii's nature reserve system: none of the islands' national parks were located in areas with significant populations of endangered forest birds. The technique identified a number of sites that had high levels of endangered forest bird species, and negotiations were then initiated to have these sites classified as national parks. Eventually the 6693-hectare Hakalau Forest National Wildlife Refuge was established on the island of Hawaii in one of the areas of highest species richness.

Gap Analysis is not only powerful—it is inexpensive. Where data on the distribution of species and preserve boundaries are available, Scott estimates that a Gap Analysis for the United States could cost as little as 1 or 2 cents per square mile.

In theory, it should be possible to use a Gap analysis to negotiate land use with developers or others interested in exploiting the land. Let us assume we have been asked to set up a nature reserve system in western Australia, within a large area about to be invaded by a team of developers who plan to create a series of small towns surrounded by farmland. Teams of ecologists have spent the last year surveying the fauna and flora of the area; we now have to determine, before the developers move in, which portions of that area should be set aside as nature reserves. If our Gap Analysis is to prove a successful tool for maximizing the amount of biodiversity we save, Bob Pressey, of the New South Wales National Parks and Wildlife Service, and colleagues from CSIRO and the Biodiversity Programme at the Natural History Museum in London suggest we have to consider our reserve system from three perspectives. complementarity, flexibility, and irreplaceability.

After we have obtained information on the geographical distribution of species, we are ready to begin selecting land for the reserve system. Our goal is to preserve a set of habitat patches containing a high proportion of the features of the original habitat. As we most likely have only a limited budget with which to buy land, our first step is to assess the plant and animal life of any existing reserves and then determine which areas of habitat complement our present pool of preserved species. At this point the taxonomic uniqueness classification system described in Chapter 5 applies: we can use it to categorize the biodiversity for each taxonomic group in each site. We than add sites sequentially that complement those already in our reserve system. This technique minimizes the risk of reaching a ceiling of acceptable reserve area before we have protected all the essential features of the original habitat. It will probably create a system of several moderately sized or small reserves, rather than a few large ones, partly because large reserves tend to overrepresent some features of an area, and partly because many plant and invertebrate species will be endemic to comparatively restricted areas.

If we examine data for our reserve system in western Australia, we find that a minimum of five sites is required to protect a significant proportion of the biodiversity. However, as is often the case, there is more than one combination of the minimum number of sites that protects the minimum acceptable level of biodiversity (in this case, there are seven possible combinations). Once we exceed the minimum number of sites, the number of combinations of sites may increase exponentially. We thus have the flexibility to choose between sites: we can trade-off their value as complementary reserves against their value if developed or otherwise exploited.

A final component has to be considered when comparing sites and that is the irreplaceability of any particular site. The more frequently a site appears in combinations, the more irreplaceable it is—the frequently appearing sites are the ones that contain the largest diversity of species not found elsewhere. Although the combinations that must be analyzed to estimate irreplaceability can rapidly become enormous for large reserve systems, sensible approximations have been developed for small systems. For these systems, we can produce maps ranking patches of habitat according to their irreplaceability. We can then determine which combination of sites will maximize biodiversity within the reserve system, while also minimizing conflicts with other land uses, such as agriculture or urban development.

Wildlife Corridors

There are very few large areas of wilderness left to be set aside as reserves, so we must find a way to make small areas do the job. We would be especially interested if there were some way that adding a small piece of land could have the effect of greatly enlarging the reserve. That may sound impossible, but one promising way to accomplish just that is to use Gap Analysis to identify corridors of land that can link together existing nature reserves. These corridors may offer the only routes through which species can undertake their annual migrations between their summer breeding grounds and their winter feeding range. Too, they can bring new individuals to repopulate a patch of habitat from which a species has disappeared. They also provide pathways for immature individuals to move between subpopulations living in patches that might otherwise be completely isolated. This influx of "new blood" can increase the effective population size of an isolated population, so that it can maintain higher levels of genetic diversity. On the other hand, the pathogens and predators that pose a threat to endangered species can travel as easily through a corridor as their prey, this could severely disrupt previously successful conservation programs.

Despite the risks posed by corridors, it is undeniable that they can considerably enhance the conservation value of smaller reserves, like most of those in North America and East Africa, by connecting them to the few large parks. At present attempts are being undertaken to make this happen (indeed, some of the money you spent on this book will go to lease land in northern Tanzania that will help link Kilimanjaro, Arusha, and Amboseli national parks). Yellowstone National Park is becoming ever more isolated from other parks in the region. Its populations of grizzly bears and other large mammals and birds may soon be endangered unless the park is connected to other parks and wilderness areas in the region. At present several conservation organizations are examining ways of maintaining the increasingly thin corridors that link Yellowstone to the large wilderness areas in central Idaho and northern Montana. It was along one of these links that wolves first managed to recolonize the Yellowstone after the so-called magic-pack migrated southwest out of Glacier National Park in the mid-1980s. In just over six years, this pack colonized several parts of central Idaho where they had previously been driven extinct. A member of the pack was first spotted in Yellowstone in 1991 by a startled wildlife photographer who realized that the large black coyote he was filming was a wolf. Once the wolves had colonized Yellowstone naturally, the National Park Service moved ahead quickly to translocate whole packs of wolves back into the park.

▶ A pair of wolves translocated from outside the park are carried in crates to their point of release in Yellowstone National Park.

One of the first litters produced by the translocated wolves was born just to the northeast of Yellowstone, on the Beartooth plateau at the edge of the Absaroka wilderness area. (The male wolf was shot by a local around the time the female gave birth to her litter.) Within 10 miles of the site where the cubs were born, a large mining company is planning to develop a sizeable gold mine. Huge amounts of rock will be moved from Henderson mountain, 4 tons for every ounce of gold, and these will be dumped on the wetlands that separate Yellowstone and the Absaroka wilderness area. Pollution and runoff from the mine will flow into Clarks Fork, a tributary of the Yellowstone river that afterward flows into the park. The mine is a fine example of "focus profits and disperse costs": it will considerably improve the welfare of a few people, while creating environmental problems that could last for a hundred years and bring distress to many of the species that inhabit Yellowstone. Plainly conservation isn't over once we have set up a nature reserve, for we then have to consider how we can manage it and protect it from damage from external forces.

Managing a Wildlife Reserve

Eureka! and Huzzah! This is a great and momentous day in the modern history of Point Lobos Reserve. While on patrol to Gibson Beach I saw a Sea Otter playing off Mr. Kellogg's property at approximately 3:30 p.m. There were at least one, maybe two, coming into the Reserve itself.

A thrilling event for me, and with great possibilities for the Reserve. For with the animals that close and the food supply available here, we could within a reasonable amount of time, have a flourishing colony here in the Reserve.

Jim Whitehead
Point Lobos Superintendent
February 20, 1954

◄ A California sea otter floats on its back among the kelp. By eating sea urchins, and thereby creating healthier kelp forests that can shelter more fish, the sea otter can alter the composition of an entire community.

Four miles south of the Monterey peninsula, on the coast of California, is the wooded promontory called Point Lobos. It is a jewel of a nature reserve, a complex system of coves, inlets, and rocky islands where the visitor can see harbor seals, sea lions, sea otters, and whole colonies of seabirds. In the wooded hills back from the shoreline, the patient observer can spot mule deer, the occasional bobcat, a large variety of bird species, and a great diversity of wild flowers that are in bloom for much of the year in the mild sea air. A restored whaling cottage stands in Whalers cove; from there boats used to sail to intercept the gray and blue whales as the animals passed along their traditional migratory routes within a mile of the California coast. Shell middens at various natural observation points along the cliffs mark the sites where for many centuries Native Americans gathered to watch the ocean and to fish or hunt for their daily meals.

There are many exquisite places in Point Lobos where one feels almost completely isolated from the frantic pace of the late twentieth century. Yet no location in the reserve is more than a mile away from California's main coastal highway, the road on which Jack Kerouac hitched lifts when returning to his cabin at Big Sur, thirty miles down the coast at what is now the southern limit of the sea otter's range. Along the California coast sea otters have acquired a mystique to rival Kerouac's own, and created an economic industry as well. It's almost impossible to find a shop in the nearby towns of Carmel or Monterey that doesn't sell seas otter postcards, sea otter castings, or books about sea otter natural history. Yet the affection felt toward these creatures is not shared by the local fishermen, who hold the otters responsible for the decline of the area's fisheries. In this view, however, they appear to be mistaken.

Otters do seem to have a dramatic impact on the structure of nearshore communities, but their presence seems to actually *increase* fish numbers. By the end of last century fur trappers had reduced sea otters to near extinction throughout much of their range. As a consequence, it is now possible to find pairs of similar sites along the Pacific coast as far north as Alaska, in which one site has no otters, and the other site has an otter population that is beginning to recover in numbers. The nearshore communities at such pairs of sites differ strikingly, demonstrating that the recovery of the sea otter over the last forty years has influenced biodiversity as dramatically as tourism has influenced the local economy.

Sea otters feed almost exclusively on sea urchins, which in turn feed on aquatic vegetation, particularly the kelp that creates spectacular underwater forests all along the Pacific coast of the United States, Canada, and the Aleutian islands that run for over a thousand miles west from the tip of Alaska. After the sea otters were hunted to near extinction, the sea urchin population flourished. Jim Estes and his colleagues at the U.S. Fish and Wildlife Service have documented that wherever sea otters become reestablished they dramatically reduce the local sea urchin population, to the benefit of kelp and other forms of aquatic vegetation on which sea

▲ Divers exploring a giant kelp forest near San Clemente island, off the coast of California. The area's dense kelp forests create a protective habitat for numerous species of fish. The individual fronds of giant kelp can grow up to six inches a day. The underwater forests they create are the aquatic equivalents of rainforests.

urchins feed. Once the sea urchins are on the decline, dense underwater kelp forests form in the intertidal zone. The kelp forms a protective habitat for the adult and larval stages of fish species, both nearshore and offshore types, and the fish attract harbor seals. This chain of events has been traced in the Aleutian islands by documenting dramatic shifts in the diet of the Aleuts who first colonized these islands. Analysis of the middens left by the Aleuts shows that sea otter remains are usually found in conjunction with the remains of fish and harbor seals. When sea otter numbers decline, because of harassment and overexploitation, the middens become dominated by the shells of sea urchins and limpets. The presence, or absence, of sea otters can cause the whole structure of the community to shift dramatically between two different states.

Sea otters are considered a "keystone" species in the intertidal kelp forest because their presence can significantly change the population densities of other species, and alter how the ecosystem functions. The dramatic impact that even a single species can have on an ecological community suggests the problems that may beset a nature reserve if a native species is lost, or an alien species invades. Without a good understanding of species interactions, park managers attempting to rectify these problems, by reintroducing an extinct species or eliminating an alien pest species, risk throwing the system out of kilter and dramatically altering the species composition in the reserve.

In this chapter we will examine the problems faced by managers of national parks in North America, East Africa, and Costa Rica. Each example illustrates how the behavior of natural systems becomes steadily more complex and harder to predict when we are dealing with a multitude of species interacting with one another in numerous ways: as predator and prey, as competitors for the same resources, or as mutualists that depend on each other to complete crucial parts of their life cycles, such as pollination or dispersal. Physicists who have tired of the relative simplicity of the "rocket sciences"—in which the emergent properties of physical or chemical systems can be predicted from an understanding of only a handful of fundamental particles—have begun to find the complexities of ecological systems to be a rewarding challenge.

The new insights emerging from quantitative studies are fast confirming the long-held suspicions of ecologists and national park managers: once an area of land is set aside as a nature reserve, we then have to learn to manage it in such a way that it maintains its distinctive biological features. Occasionally, managers will operate some aspect of reserve management as a controlled experiment, to learn how different forces structure the community. More frequently, they will rely on detailed long-term monitoring of the species in the park to tell them something about how the system functions.

▶ The changing Aleut diet reveals when during the past centuries food species were common and when they were rare. Animal remains uncovered in an Aleut midden at Amchitka island bear evidence of the sea otter's role as a keystone species in its ecosystem. When sea otters were rare for several centuries early in the millenium, the populations of sea urchins and limpets soared. Once the sea otter population had recovered, the limpet population crashed and the number of sea urchins declined markedly as well.

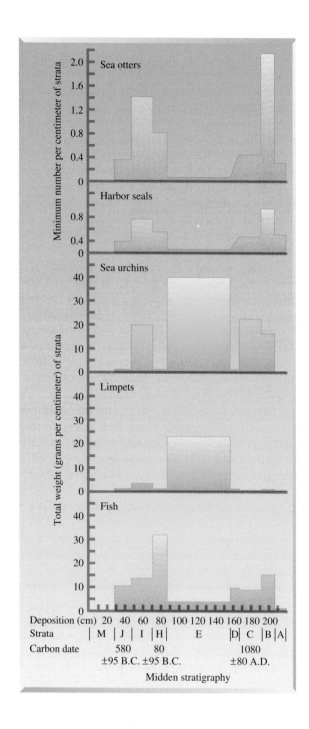

The formal IUCN definition of a national park is a place where the "ecosystem function is not materially altered by human exploitation and occupation," where the park is protected by the "highest competent authority of the country," and where visitors are allowed "for inspirational, educative, cultural, and recreative purposes." The ultimate aim of park management is to maintain the integrity of the park at a variety of levels: from the entire landscape and whole ecosystems, through communities, species, and populations, to the individual. Unfortunately, ecologists have tended to concentrate their research at the less diverse end of this spectrum. Most ecological studies have looked at the interactions between only one or two species, on a spatial scale that is usually measured in square meters rather than the many square miles required by naturally functioning ecosystems. The small scale of these studies has seemed necessary partly because funding for large-scale ecological projects has been limited, but partly because the interactions between even two species can produce a range of complex outcomes, including long-term cycles in abundance, occasional extinctions, and natural recolonization. Periods of stability are observed only occasionally and then usually for only short periods of time.

Food Webs, Diversity, and Stability

One of the first things we learn about in high school biology classes are food webs, the "flow charts" that show who eats whom in an ecological community. Food webs illustrate how primary producers at the base of the web, usually plants, are connected to (eaten by) primary consumers, or herbivores, and through them to the secondary and tertiary consumers, or predators, at the top of the web. I have already illustrated how the presence of sea otters dramatically alters the structure of their food web by reducing the density of sea urchins, the dominant herbivore species with the most significant impact on vegetation. Species may be regulated in their abundance from the bottom up, by limitations on the resources they eat, or from the top down, by the predators, parasites, or herbivores that feed on them. Because all the species in a national park are coupled together in a food web, or even several food webs if the park contains a variety of habitats, then an understanding of food web dynamics must lie, even if only subconsciously, behind the long-term management of parks and wildlife reserves.

Food webs have only recently received the attention of quantitative ecologists. Joel Cohen and F. Briand at Rockefeller University have compiled and analyzed a catalogue of 113 food webs from a wide variety of natural environments: 55 from continental settings (23 terrestrial and 32 aquatic), plus 45 from coastal settings and 13 from the ocean. Although at first sight these food webs appear as unlike as the webs of different spider species, a closer look reveals some important regularities. On average, any one species interacts with around three to five others, either as predator or as prey, and the number of these interactions in a relatively unchanging environment is slightly higher (4.6) than the number in a fluctuating environment. Another consistent pattern has around 19 percent of species at the

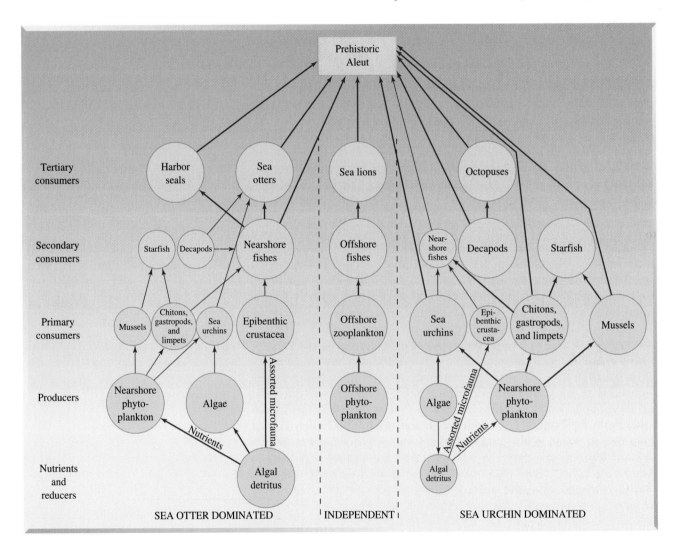

▲ This food web for the nearshore community of the western Aleutian islands shows the state of the community in the presence of sea otters (left) and in their absence after overhunting by the Aleuts (right). The sizes of the circles indicate differences between the two states in the sizes of populations; the heavy lines between circles indicate the importance of the connection. (Epibenthic crustacea are crustacea that live on the ocean bottom; decapods include squid and shrimp.)

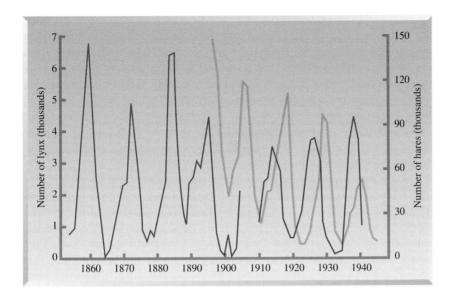

We can observe fluctuations in the lynx and snowshoe hare populations of southeastern Canada by plotting the number of furs retrieved by trappers of the Hudson's Bay Company. Rises and falls in the hare population are shortly followed by rises and falls in the number of lynx.

base of the web, 53 percent at intermediate levels, and 29 percent at the top, as "top" predators.

When a disturbance such as an invasion, a disease outbreak, or human overexploitation causes one species to rise or fall in numbers, the effects may spread in a cascade throughout the web (as they do when we break a section of a spider's web). The story of the sea otter, for example, showed how the elimination of a keystone predator transformed the overall structure of a food web: when its former prey, the sea urchin, became free to multiply, the resulting devastation to kelp forests threatened an array of fish species.

Ecologists originally thought that the complexity of food webs made ecosystems resilient to such disturbances. Charles Elton, one of the intellectual grandfathers of modern ecology, cited three chains of logic which suggested to him that complex food webs were more stable then simpler ones: first he noted that agricultural monocultures were much more susceptible to pest outbreaks than more complex natural systems; second, he noted the ease with which colonizing species could wreak havoc on simple island communities; finally, he noted that the simple communities observed in arctic regions showed a greater propensity to fluctuate periodically in numbers. Such fluctuations had been observed in western Canada by the trappers of the Hudson Bay Company as early as the eighteenth century.

From 1735 to 1942, the trappers of the Hudson Bay Company kept careful records of the numbers of snowshoe hare and lynx caught in their traps. These records reveal a 10-year cycle of population fluctuations, initially interpreted as a classic predator-prey cycle: increases in the number of hares led to increases in the number of their major predator, the lynx; the numerous lynx in turn killed off much of the hare population, then their own population declined when prey were no longer abundant. Once the lynx numbers were low, the snowshoe hare population could increase again and start a new cycle. Although it is now believed more likely that the interaction between the snowshoe hares and the vegetation they feed upon causes the hare population to cycle, the oscillations in lynx abundance still reflect cycles in their main food species. Such long-term cyclic behavior may be natural for many biological populations, although it appears to be more common in arctic habitats.

Elton was inspired by this apparent lack of stability in a simple arctic food web to suggest that the greater complexity of food webs in the temperate and tropical zones tends to produce greater stability. This stability would be explained in part by redundancy: in a diverse web, some species will replicate the roles of other species. A predator species, say, could then switch to another prey if a preferred item became rare. Elton believed that natural ecosystems naturally kept themselves in balance through the presence of a range of predators and parasites, each of which is potentially capable of regulating the abundance of organisms lower in the food web.

This conveniently tidy view of food webs was eventually challenged by Robert May, then at Princeton University. May used mathematical models to show that, in sharp contrast to Elton's view, simple systems tend to be more stable than complex systems. May suggested that the key parameters determining the stability of a food web were the number of species in the web, n; the average number of links between species, or connectance, c; and the average strength of interactions between species, i. (The strength of interaction might be measured, say, by the number of hares killed on average by a single lynx.) Mathematical models of food webs described by these parameters tend to be stable only when

$$i < \frac{1}{\sqrt{nc}}$$

This result implies that stable food webs will be characterized by weak interactions between species, and that in these webs most species will only interact with a few other species.

This debate about stability and complexity creates a dilemma for park managers trying to maximize the diversity in a region. If May is correct, higher diversity is likely to produce an unstable collection of species. If Elton is correct, it will tend to produce a stable configuration. Curiously, a number of empirical studies suggest that *both* May and Elton are correct.

Studies in the grasslands of Africa's Serengeti plain by Sam Mc-Naughton of the University of Syracuse, and in the prairies of the American Midwest by David Tilman and his colleagues at the University of Minnesota, suggest that as communities increase in diversity, species tend to be subdivided into guilds within which species compete aggressively for the same resources. One guild, for example, could consist of all nectar-feeding species, and another of all species that glean insects from dead leaves. There is little interaction between species in different guilds, so the net level of connectance in the web actually declines as the number of species increases. This subdivision of a community into guilds also creates ecological redundancy. Because the species within a guild compete strongly for the same resources, the occasional decline in one species, from the action of a specific predator or pathogen, will be compensated for by the population growth of its competitors. Species that perform the same function in a web are called ecological equivalents—in some ways they reduce the effective number of species in a web, and in their ability to compete with their other ecological equivalents they may add to the resilience of the web in the face of the occasional catastrophe such as drought.

David Tilman and his colleagues have recently compared the response of different patches of prairie to a prolonged drought. Although some plant species suffered considerably, the patches with the highest species diversity sustained the least proportional declines in total mass of living tissue. As drought-susceptible species withered away, their more drought-resistant competitors took their place. It seems that Elton was right in suggesting that a more diverse community is a more stable one. Yet May was also correct in thinking that a simpler system—of fewer interactions—is more stable. Both views make sense because, thanks to guilds, a more diverse system is in some respects a simpler one.

Rinderpest in Serengeti

Driving across the plains of the Serengeti National Park in East Africa produces a feeling of immense timelessness. The Masai herding his cattle across the short grass plains around Olduvai gorge seems a direct descendant of the people who have walked here since human time began. This initial impression of a world unchanged for millennia is misleading, however; a growing body of evidence is revealing the constantly changing nature of the plant and animal communities that form the Serengeti ecosystem. Scientists have been working in the Serengeti since the pioneering work of the Grzimeks in the 1950s, when the area's first research center was established at Seronera in the middle of the park. The history of research undertaken in the Serengeti reflects the changing trends in the study of ecological processes: in the 1960s and early 1970s, studies focused

▲ For most of the year the short grass plains of Serengeti National Park in Tanzania contain only a few gazelles. When the rains arrive in spring, over one million wildebeest, zebra and their predators move in to enjoy a short harvest of abundant food resources.

on population processes and natural history; in the late 1970s and 1980s, many studies were undertaken of behavioral ecology; more recently, conservation-oriented studies that seek to link these different themes have dominated. There has also been a resurgence of interest in nutrients, the role of fire, and the communication of disease between species.

Long-term studies of the Serengeti again illustrate how a change in the numbers of even one species may have dramatic effects that cascade through the food web. Here the long-term patterns of abundance are not so much the work of the predator-prey relationships for which the park is famed, but of a keystone species that is too small to be photographed.

The large predatory mammals of the Serengeti have been studied almost continuously since George Schaller's early studies of lion in the first years of the 1960s. Lions are very selective in what they eat: they need to consume 5 to 8 kilograms of meat per day, a goal they achieve best by concentrating on wildebeest and zebra. If they are living in larger groups, they will take buffalo in the absence of their preferred prey: if in smaller groups, they will take warthogs.

During the first years of study, the lion population of the Serengeti steadily grew in numbers, as did the hyaena population that feeds on the lions' leftovers. In the late 1970s, the woodlands of the Serengeti became saturated with lions, and the number living in this habitat has remained fairly constant since. Woodlands are the preferred habitat of lions, but when their numbers became excessive, some migrated out to the short grass plains. There the food supply is ephemeral, and lion numbers more erratic. Whereas the woodland lions produce young every year, the plains lions only produce cubs that survive in wetter years when game is abundant. Even though the lion and hyaena populations were expanding, this increase in predators did not produce a decline in their principal prey, the wildebeest and buffalo. On the contrary, the rising lion and hyaena numbers seem to be the direct result of a huge increase in the wildebeest population, which had grown from around a quarter of a million in the early 1960s to around a million and half by the late 1970s. Cape buffalo numbers also increased, although by a less spectacular amount, while zebra numbers have remained roughly constant.

The increases in wildebeest and buffalo followed the eradication of rinderpest, an acutely infectious viral disease deadly to their calves. This virus is the ancestor of both human measles and canine distemper. It was first introduced into Africa in the late 1880s, when the Italian army unknowingly brought infected cattle into Somalia. The resulting pandemic took ten years to spread to Cape Horn in South Africa, and it is estimated that during this time the disease killed more than 80 percent of the game species in its path. Contemporary diary accounts describe the plains as littered with carcasses, being devoured by vultures too bloated to take off.

In the years afterward, a sequence of epidemics attacked groups of ungulates at irregular intervals, and destroyed significant numbers of cattle as well. Both farmers and pastoralists who relied on cattle for their livelihoods came to perceive wild game species as a reservoir for the disease. In the 1950s the British government even discussed eradicating all the game in the Serengeti so that the plains could be converted into a massive cattle ranch to ensure a supply of cheap roast beef for England! However, the development of a cheap and effective vaccine for rinderpest allowed hope that the disease might be prevented. In fact, the widescale vaccination of cattle against rinderpest that was initiated in the late 1950s set into play a remarkable natural experiment. The vaccination campaign eventually demonstrated that cattle, not wildlife, were the reservoir of rinderpest, and that rinderpest probably operated as a keystone species in the East African grasslands.

No wild animals were vaccinated, and yet after the widespread vaccination of cattle, rinderpest became much rarer in both the wildebeest and buffalo populations. As fewer young wildebeest and buffalo were killed by the disease, there followed a steady increase in the numbers of these species. When their primary prey species became abundant, the lion and hyaena populations expanded. In contrast, the populations of zebras, which are not susceptible to rinderpest, remained constant.

▶ Each year the veterinary staff of
Ngorongoro Conservation Area spend
three months vaccinating the cattle population that surround the southern half of
the Serengeti. The vaccination scheme has
eliminated rinderpest virus from domestic
cattle and led to its disappearance from wild
animals.

More recent studies by Hans Prins at the University of Groningen
have suggested that rinderpest and other pathogens may also have a pronounced effect on vegetation, albeit an indirect one. In many of the clumps
of acacia trees growing in the Serengeti, the trees are all almost exactly one
hundred years old. In nearby Tarangire and Lake Manyara national parks
are much younger clumps of acacias. These seem to have become established when epidemic outbreaks of anthrax thinned the impala herds. Prins
suggests that the first rinderpest epidemic in the 1890s significantly reduced the number of animals grazing on acacia plants. Thus large numbers
of the plants could become established. By the time the wildebeest and buffalo populations had recovered, the plants had grown into trees that were
tall enough to escape the animals' reach.

There is one further twist to the rinderpest story. In the spring of 1994
reports began coming in from tourists and researchers of sick and convul-
sive lions. In a period of just over six months, almost a third of the lions in
the Serengeti succumbed to canine distemper. A similar outbreak of the
disease may have contributed to the earlier and more gradual decline of
Cape hunting dogs in the area. Levels of distemper are high among domestic dogs in areas surrounding the park, but distemper had not been widely

observed in lions until the 1994 outbreak. This latest episode suggests an additional link in the food web. Rinderpest is closely related to canine distemper, so much so that veterinarians can easily induce cross-immunity to one disease by innoculating a host with the vaccine for the other disease. When the rinderpest vaccine was being developed in Nairobi, the meat from test carcasses was distributed to people with pet dogs in the city, and incidents of distemper temporarily declined in the city's dog population. Something similar may have happened in the wild: when rinderpest was common in the Serengeti, predators may have innoculated themselves against distemper by feeding on wildebeest and buffalo carcasses infected with rinderpest. Once rinderpest was successfully controlled, levels of immunity to distemper would decline, and the predator population would become increasingly susceptible to a disease outbreak. The recent distemper outbreak in lions may reflect this absence of "naturally" acquired immunity.

There are now two major volumes that examine the ecology of the Serengeti ecosystem, as well as nearly twenty books that describe long-term studies of individual species. Yet there are very few other national parks for which comparable syntheses are available; a series of volumes is beginning to be produced describing different aspects of Yellowstone National Park, and several excellent books compare forests of the tropical Americas. No other scientific activity will as effectively increase our ability to manage national parks and wildlife reserves than further comprehensive syntheses of long-term studies for different natural ecosystems. These studies in turn are crucially dependent on the development of schemes to monitor the components of biodiversity in each park and upon a deeper experimental and theoretical understanding of how ecosystems work.

Invading Species—Weeds and Pathogens

Any European naturalist visiting North America for the first time is instantly struck by the familiarity of the vegetation around cities and coastal areas—most of the plant species are recent arrivals from Europe. The plants arrived in the bedding and clothes of immigrants and in the ballast and holds of ships. Thus the number of plant species introduced to the continent has been increasing steadily since the late seventeenth century. By competing with the indigenous species, these invading species may in turn lead to extinctions. For example, a recent survey of the 8 million acres that form the Florida Everglades found 488,000 acres to be infested with dense stands of *Melaleuca quenquinervia*, an Australian tree species that displaces the native vegetation. Melaleuca is particularly aggressive in competing for water, and its presence has exasperated the threat to native

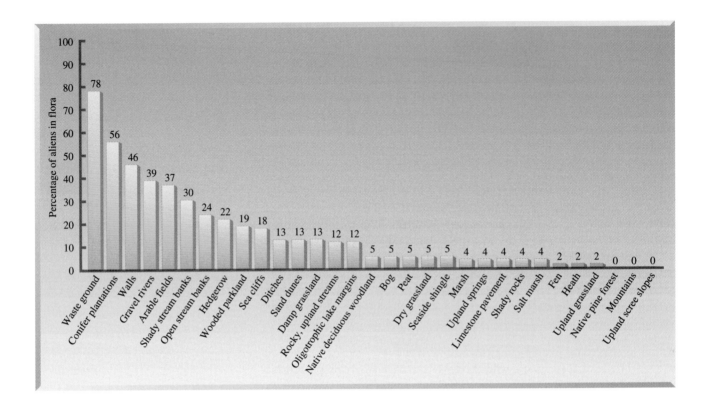

▲ The plant communities of the British Isles are ranked in this graph by the richness of their alien floras. Note that the communities richest in alien floras are those created through human activity.

plants already endangered by the increasing dryness of many inland regions of the Everglades.

Mick Crawley of Imperial College, London University, has undertaken a comprehensive survey of alien plant species in British plant communities. He has found that invading plant species tend to prefer habitats that have been disturbed by human activity, such as waste ground, arable fields, and road and railway embankments. In contrast, upland plant communities, and those on boggy or wet ground, seem more resistant to invasion. No doubt the preference showed by introduced species for disturbed habitats is partly explained by the fact that humans are more likely to unwittingly transplant seeds to places they visit often. On the other hand, wetland and upland plant communities already experience high levels of competition between their resident species and this may make them more resistant to invasion.

Isolated island plant communities seem particularly sensitive to invasion. Hawaii contains some of the most isolated islands in the world; arising from the sea as volcanos, they have never been attached to the mainland. A high proportion of the native species are endemic to Hawaii, and found nowhere else. Since the arrival of humans on the islands, the endemic plants of the lowlands have been almost completely destroyed, and replaced by sugar cane and other tropical crops. Tree species introduced from elsewhere in an attempt to encourage forestry have been particularly successful in invading the forests and excluding the native plants. For example, studies by Peter Vitousek of Stanford University have shown that myrtle trees *(Myrica faya)* introduced from the Canary islands support nodulating bacteria that are capable of fixing nitrogen. The presence of the bacteria allow the species to establish on the nutrient-poor volcanic soils of its native habitat. No tree species in Hawaii has this ability. Rather, the volcanic soils must be colonized by several communities of endemic plant species in succession before they are ready to support forests. Myrtles can short circuit this slow process; their ability to rapidly become established on bare lava effectively shuts out native species and leads to their subsequent extinction.

Introduced aliens often have an additional advantage over resident species: the predators and parasites that are their natural enemies are absent. For example, house sparrows and starlings were introduced by humans from Europe to the United States, but most of their specific pathogens and parasites did not make the trip—these were lost when sick individuals died in transit to the new habitat. Nevertheless, introduced plant species can sometimes bring along pathogens that endanger not only their hosts, but also the species that feed on those hosts. Studies in Yellowstone and other parts of the Northern Rockies have shown that to survive the winter grizzly bears rely on the large seeds of whitebark pine *(Pinus albicaulis)*, which provide half the fat the bears need to live through their overwinter hibernation. At present an introduced fungal pathogen, white pine blister fungus *(Cronartium ribicola)*, infests up to 80 percent of the pines in some parks in the Northern Rockies. In some ways, the fungus is analogous to the potato blight that produced Ireland's great famines in the mid-nineteenth century and provoked the mass migrations of the Irish to North America. Unfortunately, there are few sites available to which the grizzlies can migrate to start new lives.

Playing God with Matches

Although the population models of Chapter 3 predicted the decline of a population exposed to too many of nature's whims, in many natural communities diversity is actually enhanced by occasional disturbances. This effect was first noticed during long-term studies of invertebrates in the intertidal zones of rocky shores and of tree species growing in tropical forests. In both places occasional storms would destroy individuals in patches of shore or blow over trees in the forest, creating clearings. Species with the right adaptations, who specialize in colonizing newly opened patches of

► Although fires destroyed large areas of Yellowstone National Park in 1988, many tree species have seeds which require periods of intense heat to germinate. The burnt areas were quickly recolonized by these fire-adapted species.

habitat, would then move in. A number of mathematical and experimental studies have since showed that periodic disturbances are an important way of maintaining species diversity. The dilemma posed for managers of national parks is whether to simulate the storms and other disruptions of nature. If natural disturbances such as fire play an important role in maintaining the diversity of a habitat, should managers mimic nature by occasionally setting fire to sections of the reserve, or should concern for the safety of park visitors, or at least the quality of their visit, cause them to refrain? Recently management's response to this question was sharply put to test in Yellowstone.

In the summer of 1988, large sections of Yellowstone National Park burned in a series of fires that cast a pall of smoke across the states of Wyoming, Idaho, and Montana. Fire is one of the natural forces that determine the dynamics and composition of the lodgepole pine community in Yellowstone. In fact, lightning causes fires every summer in the park. Nevertheless, the Yellowstone fires produced an outcry from critics demanding to know what had gone wrong with the park's management policy.

The frequency and severity of the fires in any area varies with three factors: ignition, weather, and fuel. Lightening is present every summer in Yellowstone to ignite fires, but in most years the weather is too wet for extensive fires to develop. That the summer of 1988 was one of the driest summers this century had much to do with the huge scale of the fires. Fallen wood and pine litter decay slowly in Yellowstone's comparatively arid climate, so combustible material builds up between fires, and there was plenty of fuel.

Censuses taken of wildlife tell us that the fires had no large impact upon the park's major vertebrate species. Within a few years the lodgepole pine forests were beginning to regenerate in all but the areas of hottest burns, where the soil bacteria and mycorrhizae crucial to efficient root functioning were completely destroyed. Indeed, in some areas the vegetation that has sprung up after the fire has been enriched by changes in the chemical composition of the soil: nutrients such as nitrogen were volatized, while phosphorous and calcium leached from burned litter seeped back into the soil. Where this redistribution of nutrients from the forest canopy has been accompanied by the destruction of plant pathogens, the new vegetation is lusher and makes a more nutritious meal for the herbivore species that feed on it.

In the case of Yellowstone, nothing was wrong with the park's management policy. No human lives were lost in the fires, and the damage to property was minor compared to the damage created by fires that attacked the coastal forests of California in more recent summers. In fact, the effects produced by logging in the nearby Bridger and Targhee national forests are far more severe. Although some foresters argue that clear cuts simply reproduce the effects of occasional fires, their arguments ignore the fact that logging removes nutrients from the forest instead of releasing and recycling them. Furthermore, clear cuts leave neither broken branches nor snags nor standing dead trees that form nest sites for birds and insects.

◀ This aerial view shows the western boundary of Yellowstone National Park, where it abuts the Targhee National Forest. Twenty-five years of heavy logging have devastated the Targhee (on the right), which is now dominated by large clear cuts ringed by strips of standing trees.

The fires in Yellowstone epitomize the problems facing wildlife managers and resource biologists: Is it best to leave a natural park alone and assume that the species in the community will settle to some natural level of biodiversity that closely resembles that at the park's founding? Or, is it best to actively manage the park by removing or encouraging particular species? The answer to this question depends on the park and, to a large extent, on its size, since large parks require proportionately less management than small parks. Thus it is likely that laissez-faire policies will work fairly well in larger parks. However, in smaller parks species with large area requirements, predominantly predators, may become extinct, leaving their principal prey species to expand in numbers. How should park managers cope with populations that are no longer regulated by their natural enemies? Should they try to replace the predators by occasionally culling the expanding populations? Or, should they try to reintroduce the original predator species?

Controlling the Bison in Yellowstone

Since the 1970s the populations of elk and bison in Yellowstone National Park have grown markedly, in part because the annual culls practiced up until that decade were discontinued. However, the increase in bison is also a response to changes in the winter use of the park. Since the 1970s, when Yellowstone was first opened to visitors in the winter, the roads in the northern part of the park have been plowed for the use of snowmobiles. Before, deep snow had trapped the bison in two or three lower-lying valleys in the center of the park. Once these valleys were connected by plowed roads, snow was no longer a barrier to bison movement; the winter range of bison expanded considerably, and the bison population along with it. The winter operation that opens the park roads to snowmobiles and cross-country skiers, as well as to bison, brings a lot of pleasure to people who visit the park during that season. More significantly, the winter operation brings a significant amount of money into the local tourist economy (about one million dollars a month).

However, now that the older female bison have learned that a series of plowed roads lead them out to extensive winter pasture on agricultural land to the north and west of the park, they often try to lead their family groups out to feed when the forage in the park becomes overgrazed or buried by deep winter snow. The bison winter migration worries cattle ranchers who live to the north and west of Yellowstone. Bison sometimes wander through the fences enclosing ranches and start feeding on browse used for cattle. Unfortunately, some of the Yellowstone bison are infected with *Brucella abortus*, the bacteria that causes brucellosis in cattle. This disease was introduced to the bison herd at the beginning of the century when Yellowstone's small surviving herd of less than 10 bison were kept in a small fenced area with the dairy cattle used to supply milk to park visitors and staff. Although brucellosis is only likely to be transmitted in late spring when the bison have moved back into the park to have their calves, the farmers worry that bison may transmit this disease to their cattle. Were their fears to be realized, the whole of Montana, Wyoming, or Idaho could lose its brucellosis-free status, and restrictions would be put on the movement of cattle out of these states. For this reason, the cattle industry would like to have bison eliminated from Yellowstone. As brucellosis is also present in the Yellowstone elk, the removal of bison would be both futile and unethical; there are over four thousand bison and thirty thousand elk in Yellowstone—killing them all would be a major blow to the local tourist economy. Instead, tax subsidies for ranching around the park could be altered to favor ranching practices that create a "cordon sanitaire" for 30 to 50 miles of the park boundary. Furthermore, it may be more sensible to reconsider the changes in park management that have allowed the bison and elk population to expand.

The increase in elk and bison numbers may in part reflect the loss of predators, particularly the wolves and grizzly bears that feed on elk calves in the spring. At first the original predators were replaced by human hunters, but once the culls were stopped the elk population increased rapidly. Although there is much discussion about whether overgrazing and habitat change have been the result, the comparison of historical and contemporary photographs suggest that change is the rule at Yellowstone. Above all, the photographs emphasize that occasional disturbances in the form of fires or extremely harsh winters have created a mosaic of habitats in different stages of succession between open grassland and closed woodland. These disturbances are in part responsible for the high levels of diversity within the park, since a mixture of habitats can contain more diversity than a homogeneous landscape. The scale of some of the burns, which in 1988 reached 100 square kilometers in some cases, is such that maintaining Yellowstone as a functional ecosystem requires large areas of continuous habitat.

One key species that had been missing from Yellowstone for decades has recently recolonized the park. In 1988, the park had begun developing a plan to reintroduce wolves, and although local farmers had vociferously objected, public support for the establishment of an experimental wolf population was considerable. Then, in the early fall of 1993, before the park had taken any action to introduce wolves, a single black wolf was filmed by a wildlife photographer in the middle of Yellowstone. The nearest source population for natural colonists is Glacier National Park, which is nearly four hundred miles away, but wildlife biologists had noticed that wolves were beginning to migrate through the narrow corridor of parks

▲ These photographs from Pelican Cove in Yellowstone National Park, taken in October 1931 (above), July 1973 (top right), and September 1990 (right), illustrate the natural change characteristic of the park. The mid- and background of the view shows extensive reforestation after an earlier fire (perhaps in 1910).

that links Yellowstone to Glacier through the Cabinet-Yaak and Selwey-Bitteroot wilderness areas in central Idaho. Although wolves had made it to Yellowstone on their own, it was far from certain that a healthy natural population could be reestablished, since it would require luck and the gradual acquisition of all the land that links Yellowstone through to Glacier to ensure that wolves could continue to naturally colonize the park. The surest course still seemed to be for park management to bring an experimental population of wolves into the park.

Once the wolves had returned naturally, park authorities received the go-ahead to introduce an experimental population in the winter of 1994/95. Ironically, this action received the grudging support of some local ranchers, because experimental animals could be shot if they attacked domestic livestock, whereas natural colonists receive the full protection of the Endangered Species Act. The wolves were placed in three enclosures established in the northeast of the park, and once the wolves had settled down, the gates were opened and they were allowed to establish territories in the park. Early reports suggest that at least two litters of cubs have been born in the first spring, and although the father of one litter was shot outside the park, the surviving wolves are feeding almost exclusively on elk and bison calves. More introductions are planned for the next few winters until a healthy population of at least one hundred wolves has been established. The impact of wolves on the numbers of elk and bison is of course being monitored, as is their more indirect impact on the vegetation on which the elk and bison graze.

Restoring Damaged Ecosystems

Throughout the world, industrial activity and abandoned urban development have created large expanses of degraded land that is usable neither by humans nor wildlife. The danger of simply neglecting these habitats was tragically illustrated in 1966, when the forgotten and untreated waste from a coal mine slid down a hillside following a heavy rainstorm and engulfed the village school at Aberfan in south Wales. The disaster killed 144 people, 116 of them children. Similar piles of waste, as well as extensive underground holes and areas of wasteland, have been left abandoned throughout Europe, the former Eastern Bloc countries, and North America. As indus-

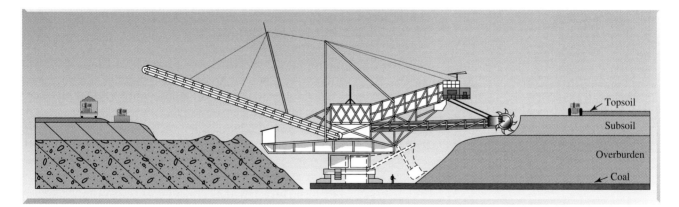

▲ The method pictured here for the surface mining of coal can completely restore high-quality agricultural land. A stripping shovel on a sort of turntable (the dashed lines in the background) scoops out the thick layer of overburden above the coal deposits, spins around, and dumps its load in the area already mined at the left. The large machine in the foreground, called a bucket wheel excavator, removes the subsoil and transfers it back to the top of the previously replaced overburden. The topsoil has already been removed and trucked back behind the excavation site, where it is replaced over subsoil smoothed by a bulldozer.

trialization proceeds in Africa, Asia, and South America, similar patches of degraded land will scar ever more of their landscape.

Restoration ecology is a relatively new discipline that seeks scientific ways of restoring degraded ecosystems. The discipline provides unique opportunities to examine how ecosystems actually work—and it illustrates how hard it is to simply re-create an ecosystem once damaged. Even if it is not always possible to restore degraded land into habitats that can be used by wildlife, it may be possible to return the land to agricultural production and thereby diminish the pressure to convert natural habitats to cropland.

Tony Bradshaw of the University of Liverpool has pointed out that the success of any attempt to restore degraded land depends on a number of factors. If the mining or other industrial activity took place in the past, and all the original soil and vegetation were lost, then restoration will be much more difficult than if carried out at the same time that the (short-term) industrial activity is actually taking place. For example, agricultural land can be surface-mined for underlying coal deposits, and yet preserved, by moving the topsoil and underlying subsoil sequentially across the strip of land being excavated. However, while there is an obvious economic advantage to maintaining agricultural land in a productive state, the financial incentive to restore wilderness areas overlying mineral deposits is usually missing. Furthermore, the highly diverse vegetation in a wilderness area cannot be expected to return instantly to its previous undisturbed state. Although some plant species may be present as seeds in topsoil that has been saved and restored, these will tend to be weedy opportunistic species. Many shrubs and trees will have to be actively reintroduced. If, as is common, the topsoil becomes degraded and compacted when temporarily removed, it will require artificial aeration and supplements of fertilizer in the early years of restoration—nitrogen-fixing plants may have to be sown into the soil to help restore lost nutrients. When ecologists compared the quality of topsoils on 46 different sites in land restoration schemes in the northwest of England, they found a huge range in the soils' ability to produce crops. None of the soils were within 70 percent as productive as garden soil, half were no better than the waste produced from brick making, and one would grow no plants at all.

The alternative approach to active restoration is called ecological recovery. This approach takes a laissez-faire attitude and assumes that the land will regenerate at its own rate. Leaving land to recover on its own is occasionally successful: Hampstead Heath, which is only 7 kilometers from the center of London, is a "mini-wilderness" that has slowly recovered from the mass of sand and gravel pits that filled the site in the nineteenth century. Laissez-faire approaches can only succeed if all the land's original plant and animal species are still present in the vicinity and able to recolonize the site at a natural rate, without competition from alien species.

When these conditions are not met, the degraded habitat will develop into a different sort of community than was present before.

Attempts to restore wetland communities in southern California illustrate some of the successes and pitfalls of restoration ecology. Almost 85 percent of the original salt marshes along the coast near San Diego have been lost as the city has expanded. Species native to the area have suffered large reductions in population size, and one plant, 7 invertebrates, 2 reptiles, and 14 bird species are now listed as "threatened" or "endangered." Joy Zedler of San Diego State University has been developing a program to restore two salt marshes as habitat for the light-footed clapper rail *(Rallus longirostris levipes)*, a federally registered endangered species. Establishing a salt marsh that is acceptable to the clapper rails has proved a formidable task. The vegetation on the artificial marshes, predominantly cordgrass, is considerably less abundant than on natural salt marshes, so when the marsh periodically floods, the vegetation is completely covered and the birds have no place to hide from predators.

The origin of the problem seems to be the sandy soils in the restored marshes. These soils do not retain nitrogen and thus cannot support plants at high density. There are consequently not enough plants to sustain large numbers of the beetle *(Coleomegilla fusciliabris)* that consumes the scale insects *(Haliapsis spartina)* that are a major herbivore of cordgrass. In the absence of beetles, outbreaks of these scale insects further reduce the growth of cordgrass. Although artificial nitrogen supplements have been applied, these have to be added fairly frequently until the cordgrass becomes sufficiently dense that a litter layer can build up able to withstand the occasional floods. As yet no clapper rails are living in either of the restored marshes. It is somewhat disconcerting that we can feel proud and awed by such technological achievements as the cruise missile and the personal computer, yet we cannot reconstruct a salt marsh—especially since a considerable amount of the marine life that much of the world relies on for food requires salt marshes and estuaries as spawning grounds.

▲ The light-footed clapper rail, such as this example of the bird photographed in California's Tijuana estuary, lives in coastal salt marshes threatened by urbanization Attempts to restore habitat for this species illustrate how hard it is to reconstruct a natural habitat.

Growing a National Park—Guanacaste

Although the world's tropical rainforests may possess the highest concentration of biodiversity, it is the world's tropical dry forests that are disappearing at the fastest rate. When the Spaniards arrived in the Western Hemisphere five hundred years ago, there were approximately half a million square kilometers of dry forest on the Pacific coast of Mexico and

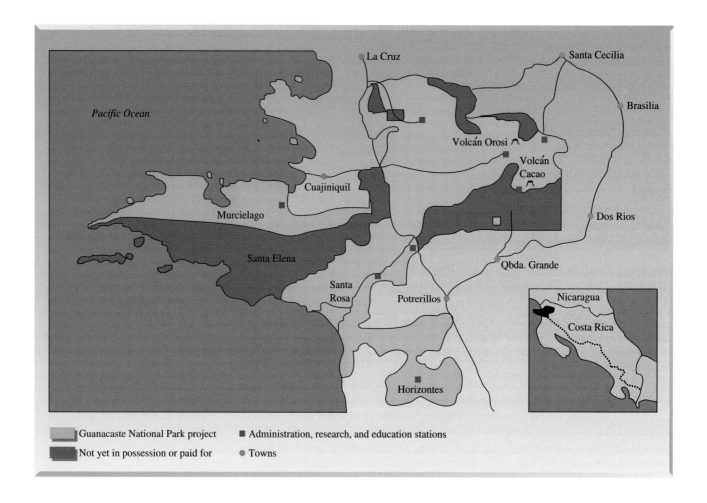

Pacific Ocean

La Cruz

Santa Cecilia

Brasilia

Volcán Orosi

Volcán
Cacao

Cuajiniquil

Murcielago

Dos Rios

Santa Elena

Qbda. Grande

Santa
Rosa

Potrerillos

Horizontes

Nicaragua

Costa Rica

Guanacaste National Park project   ■ Administration, research, and education stations

Not yet in possession or paid for   ● Towns

▲ Guanacaste National Park, where Daniel Janzen and his colleagues are restoring the dry forest, is located on the Pacific coast in the northwest of Costa Rica.

Central America. Today, less than 2 percent of this forest is sufficiently intact to attract the attention of conservation biologists. If the dry forest is to retain a true representation of the species that were there a mere five hundred years ago, it seems likely that we will have to regrow serious proportions of this habitat.

The cause of the severe habitat loss is straightforward: dry forest is much easier to burn than wet tropical forest, and fire is hard to control. Thus tropical dry forest has been cleared much more readily by settlers, especially along the coasts. Moreover, the rich soil of tropical dry forests usually makes more productive agricultural land than does rainforest soil. In general, the diversity of plant life in tropical dry forests, while as much as 50 percent less than that of adjacent rainforest, is still impressive, and the diversity of insects and other invertebrate life tends to equal that of rain-forest.

Throughout the tropics, deciduous dry forests have been replaced by cotton or sorghum fields and by cattle ranches. Restoring these habitats, and conserving their diversity of life is the challenge that ecologist Daniel Janzen, of the University of Pennsylvania, has set for himself and others. Janzen has begun his efforts at Guanacaste National Park, a 107,000-hectare nature reserve set between the coasts and mountains of northwestern Costa Rica. Much of the park's original dry forest has been destroyed and replaced by grass species introduced from Africa, interspersed with shrubby thickets.

Fire suppression is a key goal of the restoration regime. In Guanacaste, fires are not an important natural disturbance, as they are in Yellowstone. They have mainly been set by farmers and cattle ranchers to prevent tree encroachment and to provide a green flush of vegetation on which cattle may feed. If fires can be stopped at Guanacaste, trees may be able to regenerate from the few remaining small patches of native woodland. One fire-prevention measure is the frequent cutting of the introduced grass species; these species, which can grow up to 2 meters high, tend to produce very hot fires that spread easily into surrounding areas of woodland. Underused or unnecessary roads have been closed, and camping and other potentially pyrotechnic activities have been restricted to areas of the park where accidental fires can be quickly controlled. Even as few as eight fire-free years have proved sufficient to allow early second-growth forest to establish. In the continued absence of fire, this forest can prevent the reestablishment of invasive grasses.

After removing the major obstacle to restoration, the next step is to help the newly regenerating forest along by arranging for the dispersal of seeds from native trees. This step is especially necessary because many of the natural dispersal agents are missing. A highly cost-effective method of seed dispersal has been worked out: horses are fed with meal containing the seeds of important forest trees, such as the "Guanacaste" tree. The horses are then allowed to wander in the surrounding pasture, where they deposit the seeds in rich piles of manure. Once the seedlings mature into small plants, they become attractive foraging posts and nest sites for birds, and the birds in turn deposit seeds (and another appropriate nutrient supply) in their vicinity.

The work at Guanacaste illustrates two important features of a successful restoration project. First, the restoration is undertaken before surrounding fragments of the original habitat are lost. These fragments are crucial sources of seeds and of wildlife from which to restock the restored habitat. Second, once the natural history of the habitat has been understood, then subtle, but simple and cost-effective techniques are developed that mimic processes that took place in the original habitat. Restoration

▶ The human population in the seven districts to the west of the Serengeti is growing rapidly. The circles represent the average annual percentage rate of population change in the years 1978 through 1988. Each circle shows the population change within one square of a grid made of 5-kilometer squares.

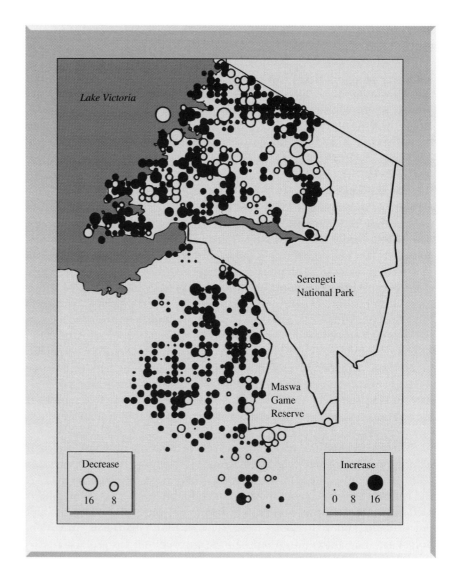

ecology is psychologically critical to conservation biologists. Instead of placing scientists in a position where we are seen to be stopping other forms of development, it allows us to actively create something new.

Even the most subtle land restoration project or park protection scheme will be doomed to failure unless the people who live in the surrounding area believe that they will receive tangible benefits. In some cases restoration projects will provide jobs for the local community, as has happened at Guanacaste. Government subsidies could even be provided to create restoration jobs. It seems ridiculous that there are many national and international subsidies for habitat destruction and yet almost none for habitat reconstruction, particularly when restored habitats will have long-lasting benefits for many people.

In other cases the presence of the park may act as an economic stimulus encouraging the growth of the human population while it simultaneously acts as a constraint on that population's ability to spread. This dangerous situation is developing to the west of Serengeti National Park, where the human population has increased tremendously in the last thirty years. At the same time that scientists were "discovering" the Serengeti, so were wildlife photographers and filmmakers—the huge spatial scale of the Serengeti and the sharp translucent light has inspired the production of many spectacular films and magazine articles. As the fame of the area has spread, it has developed into a major tourist attraction, and a source of income for the many nearby villages. The expanding human numbers in these villages are responsible for the high levels at which wildebeest, buffalo, and eland are poached for their meat, and elephant for their tusks. Unfortunately, although the number of arrests for poaching continues to increase, the number of park staff has remained roughly constant for the last thirty years. Rhinos have been hunted to extinction, and several other populations, particularly buffalo, are in significant decline. Events like

◄ A black rhino in Kenya's Nairobi National Park stands within sight of the city's skyline.

these have inspired a new recurring theme of scientific research: the impact on parks of human beings in their many guises as pastoralists, poachers, researchers, and tourists.

The Human Impact

The invading species most challenging to the managers of national parks is *Homo sapiens*. In the United States, the National Park Service recorded 258 million visitors to parks in 1990. From a purely financial perspective, increasing the attraction of parks to visitors is an important goal of park management. However, an excess of visitors can easily have as dramatic an impact as the most pernicious fungal or bacterial pathogen. Recreational users may trample rare plants or displace animals from foraging and nesting sites. Furthermore, human visitors are not always a financial bonus; many U.S. national parks have to spend up to 95 percent of their annual budgets ensuring that parks are safe for their visitors and that the park service is not legally responsible for any accidents that might befall them. Although making parks safe for people is a tremendous drain on the financial resources of the park system, it does have the positive effect of concentrating visitor activity into a small portion of each park. An important difference between national parks in the United States and many European parks is that large areas of U.S. parks remain essentially unvisited by human beings. Damage to the ecosystem is minimized, and many species are able to continue their activities undisturbed by human influence.

Any national park must find a balance between visitor use and appreciation and the potential damage that an excess of visitors can inflict. One solution is to set aside different areas of the park for different levels of use. Humans may be completely excluded from core areas of the park, while trails and camp sites are concentrated in a few restricted areas. Most important is that a buildup of human habitation around the park not cut it off from the surrounding habitat.

The most direct way in which humans influence a park habitat is through the construction of roads. The road maps of several national parks in the United States resemble food webs in their number of links. Unfortunately, these roads may form barriers to the dispersal of some species and will tend to fragment the park into small areas that mimic the larger fragmentation of the landscape outside the park. Furthermore, a poorly positioned road can cut off a herd from its water supply or scar a beautiful landscape. Denali National Park in Alaska has actually prohibited private vehicles; visitors travel to hiking trails in buses driven by park personnel. As the traffic jams in Yellowstone and Yosemite get longer each summer, this approach becomes more and more tempting.

Death Valley
National Monument

Crater Lake
National Park

Yellowstone
National Park

▲ Paved roads subdivide these national parks into fragments. The parks are not shown to the same scale.

No park is an island. Most park boundaries were set for political and historical reasons; very few capture a regional ecosystem in its entirety. In many cases, the survival of wildlife inside the park depends on the mainte- nance of species and communities in surrounding areas outside. Perhaps most important, the water and air quality within a park are the result of processes working on a much larger geographical scale. The next chapter considers how the growing human population is altering the climate and atmosphere in ways that could dramatically transform the earth's biodiver- sity, inside and outside parks.

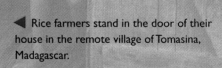

Biodiversity in a Changing World

It's early morning in a forest in eastern Madagascar. Mist surrounds the precarious-looking footbridge that crosses a waterfall on the path into the forest. Three men carrying machetes, wearing sarongs made from pieces of old sacking and tired T-shirts passed on by tourists, cross the bridge and begin sharpening their machetes on the rocks that form its foundation. When they've finished, they wander into the forest to cut palm roots, which they'll roughly chop into potted-plant holders to sell to tourists from their roadside stalls. Each plant holder sells for about 80 cents, and on a good day the men will manage to collect six to eight roots. The ferns are completely destroyed by this operation, so each day the men must wander farther into the forest to find plants of a suitable size. Eventually, they'll take the forest's last fern, and their families will no longer have a source of cash to supplement the crops they raise on their small, near-vertical patches of cleared land.

◀ Rice farmers stand in the door of their house in the remote village of Tomasina, Madagascar.

211

The children in the small group of huts that form the village are often sick. Parasitic worms are a common ailment, as are runny noses, diarrhoea, and a host of viral and protozoan infections. Many of the children won't survive to age ten. We often forget that the majority of the world's people still suffer from a wide range of infectious diseases and intestinal parasites which can sap motivation and lower economic productivity. Some plants in tropical forests have medicinal properties that could considerably reduce the impact of infectious disease (as well as the noninfectious diseases more common in the developed world); but these plant species are more likely to be exploited for their ornamental value, or simply used as firewood.

Scenes similar to the one described above are played out every day all over the world: they illustrate the simple, but underemphasized, relationship between an increasing human population and the destruction of biodiversity. Nor is the problem simply one of the world's poverty-stricken using the planet's unexplored biological resources to survive. Evidence is beginning to accumulate that economic development, in addition to human population growth, is having a significant impact on the way that many of the planet's life support systems operate. The global changes that are predicted to result from atmospheric pollution and from changes in the global nitrogen cycle pose an especially dangerous threat to natural ecosystems and the biodiversity they support. The buildup of greenhouse gases in the atmosphere and the increases in atmospheric nitrogen are direct consequences of human industrial and agricultural activity over the last two hundred years. These large-scale global changes threaten biodiversity both directly and through their assault on the general health of the planet.

Human Population Growth

In the last decade of the twentieth century, the human population will grow by just over 800 million people—a number roughly equal to the total world population in 1750. Although the annual rate of increase is now 1.6 percent, down from the peak of 2.1 percent seen in the late 1960s, we still add 93 million people each year to the number of people using the planet's resources, a number roughly equal to the population of Mexico. This many people require an area approximately the size of Belgium to supply the crops and firewood they need to survive. The most recent projections, presented at the United Nations Conference on Population and Development in Cairo during August 1994, suggest that the human population, which stands at 5.7 billion now, should stabilize at around 12 billion people sometime in the middle of the twenty-second century. This extrapolation is un-

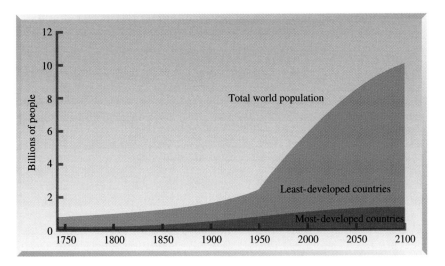

◀ Total world population has been predicted to reach 10 billion by the beginning of the twenty-second century and 12 billion by the middle of that century, with most of the increase coming from population growth in the less-developed countries.

certain because we don't know that present trends in population growth will extend into the future, and because the calculations ignore the presence of new pathogens, such as the HIV virus, that could significantly slow population growth rates.

More than two hundred years ago, Thomas Malthus pointed to the potentially huge problems created by unrestrained population growth. His arguments were based on the over-simple, but pertinent, observation that the food supply tends to increase on a linear (arithmetic) scale as more land is devoted to agriculture, while populations grow geometrically. If Malthus's logic is correct, a growing population will eventually outstrip its food resources and an increasing proportion of the population will suffer starvation. Until now, his more gloomy predictions have been confounded by the expansion of the human population into marginal lands and by massive improvements in crop productivity and agricultural efficiency, as predicted in a quotation attributed to former U.S. president George Bush, "Every human being represents hands to work, and not just another mouth to feed." Unfortunately, we are beginning to detect signs that the world's grain and fish supplies have peaked and that per capita food productivity has entered a period of decline. Our attempts to improve the yield of grain crops seem to have reached their genetic limits. Any future increase in population can only lead to a reduction in the food and land available for each person. Considering that a significant proportion of the world population is already malnourished and that loss of habitat is the major threat to biodiversity, then adding more humans to the planet is unlikely to be beneficial to ourselves or any other living organisms, with the exception of the pests, pathogens, and parasites that live off our bodies or leavings.

The human numbers in many developing countries continue to grow. Many African countries will double their populations in the next 24 years, and populations in Asia will double in 35 years. In contrast, it will take another 100 years for the U.S. population to double, while western Europe may take up to 1000 years. Because each of these doublings will effectively halve the resources available to each inhabitant of an area, standards of living throughout the world are likely to drop significantly, while malnutrition and disease spread.

The massive increase in human numbers characterizing the twentieth century is the single most important factor driving the conversion of wilderness to agricultural land. Destroying more natural habitat to create more agricultural land is not a long-term solution; when practiced by too many people, shifting cultivation and other traditional agricultural methods lead to soil erosion, the destruction of forests, flooding, and loss of biodiversity. Most of the still uncultivated land is of poor quality and will be quickly degraded. The abiding problem of too many mouths to feed will only be postponed for the short time it takes to convert the world's remaining forests to deserts. The more obvious, but politically less tractable, solution is to reduce the growth rate of *Homo sapiens.*

Between 1970 and 1990, the use of contraception worldwide increased from 30 percent of couples to 55 percent, while the average family size dropped from 4.9 children to 3.5. Aynsley Coale of Princeton University has suggested that population growth can be successfully controlled if women have the knowledge and the motivation to control fertility, and if they are given access to the appropriate methods. In Botswana, the number of children a woman is likely to have in her lifetime falls from 6 to 4 if she receives some form of education. In the Kerala district of India, where 87 percent of women are educated, the average number of children, at 2.3, is considerably lower than the national average. The motivation to control fertility is certainly present; the World Health Organization (WHO) estimates that over one-third of the 140 million women in developing countries who become pregnant each year do not want to have another baby. However, more than half these women do want another child, partly because children provide more hands to collect water, food, and firewood.

The more children that are born, the more rapidly and the more widely the available food and firewood resources are depleted. This depletion of resources takes its toll in higher infant mortality rates, which in turn encourage people to produce more children to ensure that enough survive to continue their line and support them in their old age. Julius Nyerere, former president of Tanzania, has suggested that knowing your children will survive is one of the most powerful motivations for family planning. The international childhood vaccination campaign launched by

▲ This family planning poster, distributed in villages throughout Costa Rica and Central America, contrasts the poverty of a family with many children with the better lifestyle of a planned family.

UNICEF in the 1980s may have made significant progress in reducing infant mortality from infectious diseases. However, there is often a lag between when infant mortality rates decline and when couples feel sufficiently confident to reduce the number of children in their families.

It is crucial for human health and long-term environmental stability that many orthodox religions reconsider their resistance to birth control. The Catholic Church in particular is adamant on the subject, and yet the Pope's encyclical on conception is a relatively recent historical phenomenon; the concept that every child is a wanted child was originally instigated in response to the selective sterilization programs espoused by the eugenics movement at the beginning of the twentieth century. A papal visit to any country is usually an ecological disaster: it leads to a pulse of births that simply mean more mouths to feed from fewer resources. Encouraging the concept that "every child is a wanted child" will lead only to an overliteral interpretation of "suffer the little children to come unto me." In a world with limited sources of food and an increasing diversity of infectious diseases, we can expect a growing human population to produce greater suffering among the world's most vulnerable citizens: its small children.

The most recent WHO report on human welfare, "Bridging the Gaps" (released in May of 1995), estimates that one-fifth of the world's 5.6 billion people live in extreme poverty, that almost a third of the world's children are undernourished, and that half the world's population lacks access to essential drugs. Around 12 million children under the age of five die every year from diseases for which prevention is incredibly cheap. One million die from measles alone, although it would cost only 15 cents apiece to vaccinate them. If these children were given access to clean water, sanitation, and proper nutrition, around 95 percent would survive.

The differences in life expectancies between the developed countries (around 80 years) and the poorest developing countries (around 40 years) reflect the huge inequalities in the way the world's resources are distributed. At present, one quarter of the world's population uses 75 percent of the energy consumed, 80 percent of the fossil fuel, and 85 percent of timber products. Thus, one United States citizen has a net annual impact on the environment equivalent to 150 Bengalis or 300 Malians. If we allow for differences in life expectancy between the United States and Mali, then over the course of a lifetime one U.S. citizen uses up as many nonrenewable resources as 500 Malians. These inequalities fuel resentment and hostility toward the more developed countries. It is much harder for any well-intentioned advice on natural resource management to be accepted in countries whose inhabitants perceive success to be associated with the unsustainable standards of living achieved in the United States and western Europe.

Poisoning the Environment

One way to gauge human impact on the environment is to calculate the net global amount of terrestrial annual primary production used by humans. Global net primary production, abbreviated NPP, is the amount of new material produced by plants through the photosynthetic reactions powered by solar energy. Essentially, global NPP is an estimate of the annual amount of food produced and made available for consumption to the world's consumers and decomposers. Peter Vitousek and colleagues at Stanford University have calculated that nearly 40 percent of net terrestrial

Terrestrial Net Primary Productivity (NPP) Used, Dominated, or Foregone as a Result of Human Activity	
	NPP (Pg/yr)[a]
Total terrestrial NPP	132
NPP used	
Consumed by humans	0.8
Consumed by domestic animals	2.2
Wood used by humans	2.4
Total	5.4 (4% of total)
NPP dominated	
Croplands	15
Converted pastures	10
Tree plantations	2.6
Human-occupied lands	0.4
Consumed from little-managed ecosystems	3
Land clearing	10
Total	41 (31% of total)
NPP lost to human activity	
Decreased NPP of cropland vs natural systems	10
Desertification	4.5
Human-occupied areas	2.6
Total	17 (5% of total)

[a] All figures are in petagrams (Pg) of organic material. One petagram is equal to 10^{15} grams, or 10^9 metric tons.

primary productivity is co-opted by the human population. As detailed in the table on the next page, only 4 percent of this co-opted NPP is actually consumed by human beings or their domesticated animals; the rest is accounted for by the large portion of crops, pasture, or plantations that remains unconsumed and by the plant life lost when land is cleared or degraded. A recent study of the world's fisheries estimates that we use 8 percent of global aquatic primary production, but perhaps 24 to 35 percent of the production of freshwater systems. There is thus little net primary productivity left over, either for all other living organisms or for a larger human population.

We have seen that the human population is doubling every ten years. Although we could use our food and mineral resources more efficiently, perhaps by relying more on wind and solar power, simple arithmetic tells us that twice 40 percent is 80 percent, and twice 80 percent is not possible in a finite world! Furthermore, as we reduce the resources available for other species, we quite probably hinder their ability to provide services such as nutrient and mineral recycling that are crucial to human life.

The services provided by ecosystems include maintaining the gaseous composition of the air we breathe and that of the upper atmosphere. Within the last ten years new evidence has appeared suggesting that human activities have been largely responsible for climbing levels of atmospheric carbon dioxide. As its levels increase, this gas is steadily producing a greenhouse effect that will lead to rapid changes in the earth's climate, although at present sulfurous emissions from industry mask the effects of carbon dioxide around most of the world's centers of human activity. Ironically, these sulfurous emissions have also produced acid rain, a form of pollution that has disrupted and destroyed forests in areas of Europe and North America far removed from sulfur-emitting industries.

Aside from damaging the habitat in which they live, acid rain seems to be causing direct harm to Europe's snail and bird populations. Jan Graveland and colleagues at the Netherlands Institute of Ecology have discovered that an increasing number of small woodland birds are producing eggs with thin and porous shells. In particular, the proportion of great tits (*Parus major*) laying eggs with defective shells increased from 10 percent in 1983/84 to 40 percent in 1987/88. The most-affected populations live in woodlands growing on poor soils, where snail numbers have also fallen significantly. Graveland suggests that the deposition of acid rain is causing calcium levels in these areas to decline. (It can be shown experimentally that snail numbers rise again when calcium in the form of lime is added to the soil.) Snails require calcium to produce their shells, and birds that feed on these snails obtain calcium for the shells of their eggs. Dependent on a calcium-deficient diet in the absence of snails, females produce thin-shelled eggs that crack before hatching—an effect that is reminiscent of the widespread damage done to birds of prey by pesticides in the 1950s and 1960s, so vividly described by Rachel Carson in her book *Silent Spring.*

This century's impressive gains in agricultural productivity have become increasingly dependent not upon additional human hands to work, but upon technological aids such as pesticides, fertilizers, and farm machinery. These innovations have increased standards of living for many, but their short-term benefits have to be off set against their longer-term, more widely dispersed impact on the environment. In the United States, the $3 billion annually invested in 500,000 tons of pesticides produces about $16 billion in increased crop yields. David Pimentel and his colleagues at Cornell University have attempted to quantify some of the annual additional costs of pesticide use in the United States. These costs include damage to human health ($787 million), animal poisonings and food contamination ($30 million), the destruction of beneficial natural enemies and predators ($520 million), the evolution of resistance in the pest species ($1.4 billion), losses of wild birds and fish ($2.1 billion), crop losses due to misapplication and contamination ($942 million), and groundwater contamination ($1.8 billion). The total annual additional cost amounts to $8.123 billion. When this cost is subtracted from the previous apparent profit from pesticide use, we end up with a net profit of just under $5 billion in additional crops.

In a wonderful example of the speed with which natural selection can operate, insect pests have developed genetic resistance to many of the compounds used to control them. Now new, more expensive compounds have to be developed to control strains of pests that are resistant to cheaper ones. A result is that it costs more for the farmer to produce the same crop. The development of genetic resistance is also leaving the natural enemies of pests more vulnerable to pesticides than the pests themselves. The natural predators and parasitoids that have previously controlled pest species develop resistance to insecticides at a much slower rate than the pests— partly because their generation times are slower and partly because fewer survive to leave offspring, since they accumulate larger, more concentrated doses of the compounds when feeding on the dead and dying pests treated with insecticides. Moreover, because insecticides such as DDT were very cheap when first introduced, they were applied very inefficiently under the assumption that at least some of the compound would reach its target. David Pimentel and Lois Levitan of Cornell University estimate that today only 0.1 percent of pesticides applied to crops actually reach target pests. The rest slowly finds its way into the soil, water, and air, from which it makes its way to infect species it was never intended to harm.

Species at the top of the food chain are hit hardest by pesticides. Increasing amounts of pesticides accumulate in successive species going up the food chain as each predator absorbs the pesticides present in each of its prey. In the United States and Europe, the use of pesticides significantly reduced populations of birds of prey, such as bald eagles, peregrine falcons,

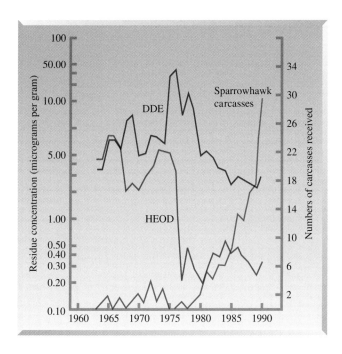

◀ More sparrowhawk carcasses are being recovered in Great Britain since the use of DDT was discontinued in the 1970s, suggesting that the sparrowhawk population has been able to recover and expand. At the same time, the levels of DDE and HEOD, two metabolites of DDT, found in the livers of these recovered carcasses have declined.

and sparrowhawks, especially during the days of DDT. As with current debates about global climate change and increases in atmospheric nitrogen, it took several years before governments acknowledged that anything was wrong. Ian Newton and his colleagues at the Institute of Terrestrial Ecology have recorded the recovery of sparrowhawk populations in Great Britain once DDT was banned in Europe.

Although DDT is now banned from use in the United States and many European countries, it is still produced in these countries and exported to farmers in Africa, Asia, and South America. Pesticide residues have now been detected in penguins and seabirds from isolated parts of the arctic and antarctic, in regions separated by thousands of miles of ocean from the nearest agricultural site where pesticides have been used. We often forget that our own species feeds at the top of most of the world's food chains; provided that significant amounts of DDT are used somewhere in the world, it will eventually appear in species that we rely on as sources of food. In particular, it will find its way into the food webs of the world's oceans from which we obtain most of our fish. A growing body of evidence suggests that infertility is on the rise both in people exposed to pesticides and in their children. We may eventually find that although the increases in agricultural production brought about by use of pesticides were initially cause for celebration, there is no free lunch until all the bills have been paid.

Distortions of the Nitrogen Cycle

Nitrogen is an essential element in animal and plant proteins. Although nitrogen gas makes up four-fifths of the atmosphere, plants cannot take up the element in its simple gaseous state. The nitrogen must first be "fixed," or converted to a usable chemical form. Many plant species have evolved mutualistic relationships with bacteria and fungi that can fix atmospheric nitrogen. Thus nitrogen exists as a large pool of atmospheric nitrogen, and in much smaller quantities, as a constituent of compounds formed by bonding with carbon, hydrogen, and oxygen. Nitrogen in its various forms is caught up in a planet-wide cycle that moves it among animals, plants, soils, sediments, and bodies of water.

Ice cores taken from the arctic and antarctic indicate that until the year 1600 nitrogen concentrations in the atmosphere remained remarkably constant at 285 parts per billion. Since then the amount of nitrogen in the atmosphere (usually in the form of nitrous oxide, NO_2) has been steadily increasing at a present rate of about 0.3 percent per year. Nowadays most of the extra nitrogen comes from the application of fertilizer, from automobile exhaust, and from leguminous crops, which are good fixers of the element.

Peter Vitousek estimates that more than half the nitrogen produced in human history has been produced since 1982! We are beginning to detect the indirect consequences of this excess of nitrogen. A number of experiments have shown that insects grow faster when feeding on plants raised in a nitrogen-enriched environment—and fast-growing insects can do more damage to the plant. Studies in the agricultural regions of the Netherlands have shown that nitrogen-demanding grasses have increased in density over the last few years, while plant species adapted to infertile soils have declined. The insects raised on these nitrogen-rich plants may well be contributing to the dieback in some European forests. In coastal zones, nitrogen deposited into rivers and estuaries has fed the growth of algal blooms that turn water green—an example of eutrophication. In parts of India and Peru these algal blooms may be responsible for a number of cholera epidemics—the pathogen can thrive and multiply in the conditions provided by eutrophication.

Take Me to the River!

As the practice of agriculture has intensified in the last one hundred years, so has the world's demand for water. The new high-yield strains of rice and other crops require much larger amounts of water than less productive crops. Locating this water is fairly straightforward where there are large underground aquifers, but it has taken many hundreds of years for these aquifers to acquire their supplies of water—and some have been drawn down in less than a decade. Of course, it may be possible to build a dam to slow the rate at which water runs away from an area, but many of the world's major rivers flow through more than one country. A dam built in one country could almost totally block the water supply of a country downstream. The Nile flows through 10 countries, and although it supplies 97 percent of Egypt's water, it also supplies water to Sudan, Kenya, Rwanda, Burundi, Uganda, Tanzania, and Zaire. The more the water needs of these countries increase, the less water will arrive in Egypt, or even Sudan—the only two of these countries to have signed a water-use treaty.

A recent report by the World Bank predicts that most of the wars fought in the next century will be about water. The worldwide demand for water is expected to double in the next 21 years, and the increase in demand will be significantly larger still in countries such as Israel, Jordan, and Syria, where aquifers are already running low. Part of the huge demand for water will stem from the expansion of agricultural activity onto drier lands that require constant irrigation. Another part will come from the growing water requirements of industrial and urban areas—even now Los Angeles imports water from more than one thousand miles away. The wholesale demand for water is already obliterating wetlands and estuaries—in many cases entire rivers are disappearing before they reach the sea. As we saw in Chapter 2, their loss threatens the biodiversity of some of the biologically richest regions in the world.

Too Darn Hot!

In the last ten years, the claret-growing regions of France have enjoyed eight of the best wine vintage years of this century. In particular, the warm summers of 1990 and 1991, following wet winters, produced some of the finest clarets ever known. If these hotter summers signal a global warming trend, wine connoisseurs in some parts of the world might regard global climate change as a potentially wonderful event. Similarly, landowners in northern Canada and Siberia might look forward to the warming of land that is permanently locked up in permafrost, a development that could make their properties important contributors to the world grain supply. Nevertheless, not all regions will gain a more productive climate: longer, drier summers may reduce the grain harvest in the American Midwest and even induce a prolonged return to the dust bowl era of the 1920s and 1930s. Many cities will endure the heat and humidity of summer for a longer period, and the increased melting of the polar ice caps could bring flooding to coastal areas throughout the world.

The intrinsic variability of the weather tends to mask systematic trends that may indicate long-term changes in climate. Nevertheless, most climatologists are now satisfied that we are entering a period of global warming, caused mainly by a buildup of carbon dioxide (CO_2) and other greenhouse gases in the atmosphere. Around 20 percent of the excess atmospheric carbon comes from fires in the Amazon and other tropical forests; the rest comes from industrial and agricultural activity, particularly the burning of fossil fuels such as coal and petroleum. To reduce present levels of atmospheric CO_2 by 50 percent would require a complete cessation of rainforest destruction and the planting of one or two million square miles of forest.

Atmospheric CO_2 is increasing at an annual rate of between 1 and 2 percent, fast enough for the amount of CO_2 in the atmosphere to double sometime in the middle of the next century. The net result would be a global mean temperature increase of between 1 and 5°C, occurring over the next fifty years. This rate of increase is around a hundred times faster than other temperature changes of the last 20 million years. To persist, many species of animals and plants will have to adapt to hotter temperatures or migrate toward the poles. The biggest worry of most conservation biologists is that the rates of climate change may be too rapid for most species to withstand, particularly when the effects of climate change are compounded by pollution, overexploitation, and habitat degradation and fragmentation. Rob Peters of the World Wildlife Fund has pointed out that, if global warming continues, the present sites of many nature reserves may soon no longer enjoy the meterological conditions necessary for the well-being of the species they were established to support. Although migration may be a simple option for most animals, it is a less viable option for plants, upon which the nourishment of most other life depends. A 1°C

▼ Left: Concentrations of CO_2 over the past 160,000 years have been determined by analyzing air bubbles trapped inside Antarctica's Vostok ice core. The recent rapid jump in concentration has taken levels of CO_2 to above 350 microliters per liter of air, distinctly higher than any concentrations of the past 160,000 years. The dashed line represents the estimated human population over the same period. Right: Global surface temperatures seem to be rising in this century, according to data recorded at land and island stations and from the ocean surface.

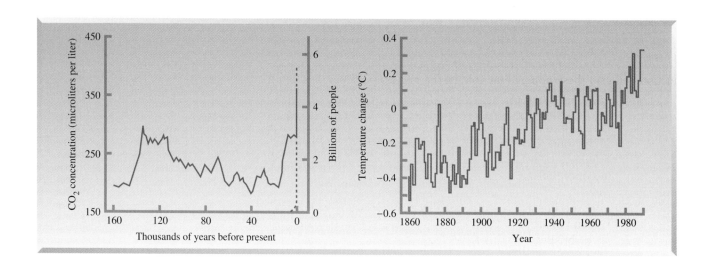

increase in temperature is equivalent to a 60 to 100 mile change in latitude. At the projected rates of climate change, trees will have to migrate at rates in excess of 500 kilometers a century (or one meter an hour)—a rate of migration impossible to achieve for most species. Studies of pollen distribution for deciduous trees have led Margaret Davis at the University of Minnesota to suggest that historical migration rates have never risen above 200 kilometers a century, with 10 to 40 kilometers the norm. Rates of migration aside, in many regions migration routes will be blocked by urban areas and agricultural land as well as by lakes, rivers, and mountain ranges. Furthermore, because different tree species will migrate at different rates, the species composition of forest communities is likely to be considerably scrambled.

If migration is a viable response to climate change, species living at higher altitudes in mountainous areas would seem to have an advantage; there is less human activity in these areas, and the species would have shorter distances to travel, since a 500-meter increase in altitude compensates for a 3°C increase in temperature. Mountains are smaller toward their peaks, however, and thus the areas of habitat available for upwardly mobile species will be smaller. Denis Murphy and Stuart Weiss of Stanford University, using the techniques of island biogeography to examine the area requirements of mammals living on mountaintops of the Great Basin region of Nevada, have estimated that 10 to 50 percent of these species would go extinct following a 5°C increase in temperature.

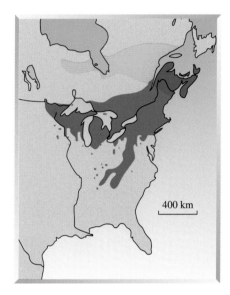

▲ According to one prediction, the geographical range of yellow birch (blue) will move considerably northward (orange) if the concentration of CO_2 doubles. Although the yellow birch's present and future ranges would still overlap (green), suggesting that the species will persist, species with smaller ranges may not be so lucky.

Genetic Responses to Climate Change

If they cannot migrate quickly enough to survive, plant and animal populations will have to adapt to new conditions right where they are. Natural selection will lead these populations toward the necessary adaptations, but a too rapid climate change may push them beyond the limits of selection. How well a population responds to selection depends on its level of genetic variability. Where such variability exists, some individuals will perform slightly better in some environmental conditions than in others. If this ability is inheritable, organisms that do better under a changed climate will tend to produce more offspring than those who were better adapted to the previous climate.

Laboratory experiments have examined the rates at which populations of fruit flies respond to artificial selection for a particular genetic trait, such as the number of bristles on their front legs. Researchers allow only the

▲ The area able to support boreal forest (green) is reduced as we go higher in the southern Snake range, in Great Basin National Park. If lower elevations became too warm for this type of forest, the area available higher up would be much smaller. The outer line is the 7500-feet contour, and successive black areas from left to right are at 8000 feet, 9000 feet, 10,000 feet, and 11,000 feet.

flies with the largest number of bristles to reproduce. This type of artificial selection usually reaches a response plateau within 20 to 30 generations. At that time all the individuals in the population end up with roughly the same number of bristles on their front legs. All the surviving individuals are descendants of the individuals in each generation having the largest number of leg bristles. Unless a mutation produces new genetic material that codes for bristle number, the response to selection ceases. Small populations, having lower levels of genetic variability, will reach a plateau sooner, after achieving less change.

Those species best adapted to cope with climate change are those able to both migrate and adapt: such species will have wide geographical ranges, large population sizes, and excellent powers of dispersal. Needless to say, few species with these characteristics are endangered or threatened. Most endangered species are likely to be restricted in their geographical range and fragmented into small populations with reduced genetic variability—characteristics that make them all the more susceptible to climate change. Global warming will present a potentially terminal challenge to the small populations of many endangered species.

Many species have evolved lower basal metabolic rates (BMR) to cope with the stress of an extreme environment; species living in deserts, for example, have lower BMRs for their body size than species living in habitats where water is more plentiful. A species that undergoes a reduction in metabolic rate must often divert resources away from growth and reproduction to maintenance and survival. The birth rate falls and population density declines, while the range of the species may shrink. Notice that all these factors tend to work against the persistence of the species, while at the same time its increased resistance to stress is working for the survival of particular individuals.

The contradictory effects of a lower metabolism have been demonstrated in laboratory experiments with the fruit fly *Drosophila melanogaster*. When laboratory flies are exposed to dry conditions in the laboratory, only those individuals most resistant to desiccation survive and produce offspring. After several generations of selection, a strain of fruit fly evolves that has lowered metabolic rates and an increased resistance to desiccation. Curiously, although these individuals show a drop in their average birth rate, as expected, they also develop greater tolerance to a range of other stresses such as starvation, heat shock, and toxic concentrations of organic compounds.

This result suggests a curious paradox for free-living populations of endangered species threatened by climate change. At the margin of a species's range, conditions are likely to be tougher, and population density will be lower (hence the local genetic variability will also be lower). These same populations, although disadvantaged in some senses, are likely to have been selected for stress resistance. In contrast, populations from the center of the range may have higher metabolic rates and genetic diversity, but lowered resistance to stress. Neither population alone will have the combination of high resistance to stress and high genetic variability that would guarantee its survival. We may have to take steps to conserve both sorts of population.

The Melting of the Polar Ice Caps

Climate changes are predicted to be more pronounced closer to the poles. The effect on the polar ice caps, which may undergo a substantial reduction in size, could well be profound. The edge of the ice barrier refracts and concentrates the sun's oblique rays, creating a massive phytoplankton bloom that grows on the underside of the ice edge. This bloom provides much of the nourishment for marine birds and animals in the polar oceans. As the polar ice caps shrink, and the perimeter of the ice edge diminishes, the area in which the bloom can grow will contract, with effects that cascade throughout arctic and antarctic food webs. Disastrous consequences could be in store for most of the world's few surviving large fisheries, which tend to be found at the boundaries of the Arctic-subarctic and Antarctic oceans.

If the warming in polar regions is as severe as predicted, then arctic tundra may completely disappear as a habitat. Should the upper 2 or 3 meters of permafrost that binds the tundra together be lost over the course of

▶ Veterinarians collect blood from the endangered Florida panther. Increased human development and loss of coastal habitat due to flooding have squeezed the range of this species into a tiny area.

the next ten to twenty years, the wet coastal tundra will also vanish. The populations of migratory birds and mammals that breed in this habitat could plummet.

A rise in sea level of 2 to 3 meters, brought to pass by the melting of the polar ice caps, could be disastrous for coastal areas, which are crucial spawning grounds for much marine biodiversity. A look at the possible effects on coral species by Carleton Ray of the University of Virginia illustrates how difficult to predict such effects can be. Corals need shallow water and plenty of sunlight. An initial analysis might suggest that fast-growing species such as *Agriporas* would be able to cope with a quickly rising sea level and remain close to the water's surface. However, as global warming alters temperature differentials between different regions, tropical storms may become more frequent. Because *Agriporas* are very susceptible to storm damage, slower-growing corals such as brain corals might actually do better than rapid growers in a warmer world.

The state of Florida in the United States may already be feeling the effects of global warming and a rising sea level. Attempts to determine the site of Columbus's landing to celebrate the quincentennial of his arrival were frustrated by the fact that the site is now offshore. Because of its low-lying topography, the Everglades National Park loses 10 square miles of area for every foot of mean sea level rise. Its endangered marshland species are thus increasingly trapped between the rising sea and the large human settlements located inland. The number one cause of mortality for the endangered Key deer, Florida panther, and manatee is collision with vehicles. The chance of being struck by a vehicle increases as populations of these species become squeezed into smaller areas bounded by the sea and human populations.

Climate Change and Interactions Between Species

Charles Park's classic experiments on competition between flour beetles undertaken in the 1940s elegantly illustrate the crucial influence of local climate on interactions between species. In his examination of two species of *Tribolium* flour beetle, Park showed that colonies of each species could exist at a broad range of temperatures and humidities when present by themselves in an experimental flour chamber. In contrast, when two species were present in the same chamber, only the better competitor would flourish. Under hot-moist conditions, *T. castaneum* was consistently the dominant competitor, while *T. confusum* dominated under cool-dry conditions. Parks' experiment hints at the complex climate-influenced interactions that are likely to exist in more diverse natural communities. Small changes in the competitive ability of a species in one part of a food web may lead to extinctions in other parts of the food web, as changes in population density are magnified by predator-prey and host-parasite interactions.

Tsetse flies provide an example of how the reaction of a single species to global warming could affect whole communities of organisms. These insects act as guardians of biodiversity in many areas of the African savanna, because as vectors of the trypanosomes, a type of protozoa that causes sleeping sickness in cattle and humans, their presence discourages farmers and pastoralists from raising cattle. Consequently, many wilderness areas are preserved for wildlife that might otherwise be converted to pasture. The crucial effect of temperature and humidity on the distribution of one tsetse species, *Glossina morsitans*, has been studied by David Rogers and Sarah Randolph of Oxford University. Their data suggest that a 2°C increase in temperature might significantly change the distribution of *G. morsitans* during the next fifty years. In particular, tsetse numbers may decline in much of central Africa: their disappearance would open up large areas for cultivation and lead to losses of biodiversity in the Central African Republic, Uganda, and some of West Africa.

How well communities are able to withstand climate change may depend on their diversity. David Tilman and colleagues have been studying the structure of grasslands in Minnesota for the last 12 years. During the years 1987 and 1988 they had the opportunity to examine the response of grassland communities to the worst recorded drought in the last half-century. Their analysis clearly shows that the diversity of species present was a key factor in determining how much of the grassland community survived the drought. Communities with fewer than five grass species lost a significant amount of living tissue (biomass). The implications are worrisome for agriculture in the Midwest. Will the monocultures common in contemporary farming survive the more frequent droughts predicted for the future? The herbicides that are used to control "weeds" reduce plant diversity in areas surrounding croplands and increase the likelihood that these areas, too, will lose biomass in future droughts. Without as many plant roots holding the soil in place, the land would become susceptible to erosion, which could in turn create a new and extensive dust bowl.

The Politics of Climate Change and Biodiversity

Climate change is not entirely a consequence of industrialization: although the burning of fossil fuels contributes 50 to 70 percent of the greenhouse effect, the destruction of tropical rainforests contributes another 20 to 30 percent. Reducing industrial pollution and lowering rates of rainforest destruction are the only ways to reduce rates of climate change. Slowing climate change is thus synonymous with constraining other processes that reduce biodiversity. Trees absorb more CO_2 than they release: the benefits we receive from forests as sinks of CO_2 far exceed the small economic benefits we receive from the harvesting of their timber or their short-term use as agricultural land. Indeed, if we want to conserve biodiversity and slow global warming we should begin very rapidly to plant new forests.

Unfortunately, the uncertainties inherent in predicting the weather encourage the hesitancy of governments. Politicians in both the developed and developing worlds are intimidated by the seemingly formidable costs of attempting to reduce global climate change. Yet global warming will pose a threat not only to biodiversity, but also to objects closer to the hearts (and wallets) of the world's least endangered species. Rising sea levels are going to do unpleasant things to the value of real estate, particularly in the coastal regions where a third of the world's population lives. Aware of the danger, politicians are at last talking about the potential costs of ignoring global warming, in Washington, Paris, Geneva, and in particular the capitals of low-lying nations, such as the Netherlands and the Caribbean island states, that risk disappearing underwater if the polar ice caps melt.

Ultimately, the human economy is not separate from the environment in which we live. Indeed, as the former senator for Colorado Tim Worth has pointed out, the global economy is a wholly owned subsidiary of the environment. There is a reciprocal interaction between the two that is fundamental to our understanding of sustainable development. Despite the dangers posed to our future health and productivity, no one is going to stop current rates of economic development by arguing that there will be significant costs sometime in the future. Instead we have to start creating alternative forms of economic development that recognize the true value of biodiversity and the services provided by the natural environment. We then have to work out more efficient ways of taxing activities that damage the environment and promoting innovative schemes that lead to more sustainable use of natural resources. The next and final chapter describes how we can begin to resolve this debate between economic development and the conservation of biodiversity by more realistically considering the value of the services supplied by natural ecosystems and by stimulating more innovative sustainable use of the products of biodiversity.

Chapter 10

The Wealth of Nature

It's easy to find the large chunky seeds of the Tonga tree on the floors of the forests of Trinidad. These Tonga beans were once the mainstay of the local economy. When boiled in alcohol they produce a liquid that was widely distributed as a perfume in the 1920s and 1930s. People believed that a whiff of this perfume made the heart beat faster! And they were right: a substance in the Tonga bean mimics mammalian neurostimulants that increase the heartbeat. The chemical was eventually isolated by a team of chemists, and an artificial analogue synthesized. Drugs modeled on this analogue now make a substantial sum of money for a large pharmaceutical company. As is the case with many cultural curiosities, the perfume is no longer fashionable, and the bean is no longer extensively harvested. Instead one can hear the occasional clunk of the axe and the distant laughter of children as small areas of forest are cleared to provide patches of cropland for Trinidad's expanding human population.

◀ The opportunity to see orcas up close brings sea kayakers, and tourist dollars, to the coast of British Columbia. Ecotourism is one way that biodiversity can provide jobs and revenue, and an incentive for its own preservation.

▲ Tonga seeds, no longer in demand, are left unharvested on the forest floor in Trinidad.

We have already seen that conservation biologists are presented with a range of scientific challenges, such as enumerating and classifying the earth's biodiversity, understanding the population dynamics of small populations, and determining how ecosystems and food webs function. From their studies they are gathering scientific knowledge that will help us to make wise decisions about how to go about preserving biodiversity. But scientific understanding alone obviously isn't enough. As we have seen in the last chapter, biodiversity will ultimately never be safe until the human population stabilizes at a reasonable level and we restrain our use of the planet's resources. But in the meantime, there are people living near forests and other natural habitats who need jobs. There are large, influential businesses that believe it is in the interest of the economy, and certainly of themselves, to develop these habitats or at least exploit their products as rapidly as possible. And there are politicians pressured by both groups and, in many cases, by their nation's desperate need for foreign currency. Everything that we have discussed so far suggests that the risks these economic and political problems pose to biodiversity must be eliminated or considerably reduced within the next ten to thirty years, or we will lose at least half the species that share the planet with us.

If conservation biology were a purely experimental science, some people might argue that letting a whole range of species go extinct could be a valid way of determining which parts of biodiversity are important. Yet this purely experimental approach is plainly flawed—it is the equivalent of saying to a sick person that you don't seem very sick, but we'll sell off your organs to pay for your hospital bills; once you get really ill, we'll work out which organs were most important and put them back, and if we can't find them we'll try to develop some way of replacing them. This is essentially the approach that most governments have adopted to managing their biological wealth and the health of the environment. The challenge presented to conservation biologists is to take what they understand scientifically and see if it suggests a policy for preserving biodiversity that is both economically viable and politically acceptable.

To achieve this goal we have to convince economists, politicians, and corporate leaders that biodiversity has three crucial features. First, biodiversity has a large potential to create new business opportunities—in particular, the expanding human population is going to require drugs to treat the infectious diseases that are a direct consequence of our crowding together into cities. Second, the truly sustainable use of biodiversity provides a huge opportunity to create new jobs—not only in labor-intensive extractive industries, but also in the management and reconstruction of nature reserves. The traditional extractive industries such as forestry, hunting, and fishing have led to major declines in biodiversity and yet the vast majority of people employed in these industries have not made a significant income from their activities. A major priority for conservation policy is to create new jobs for these people so that they can enjoy a similar quality of life while helping to reconstruct and maintain natural ecosystems.

Third and finally, biodiversity helps maintain human health and economic well-being through the undervalued but vital services provided by natural ecosystems. We have seen in the previous chapter that many human activities are beginning to cause long-term changes in climate, agricultural productivity, and the availability of water. The earth's remaining forests and oceans are the best buffer we have in slowing these large-scale changes—the tropical and arctic forests and the world's oceans are in many ways like the lungs and liver of the sick patient described above. If we optimistically think of the human population as the brain of this patient, it seems clear that continually removing portions of the lungs or poisoning the liver is probably not the best way to keep the brain alive.

Curiously, there is a fundamental similarity in the studies of ecologists and economists: both are concerned with the rates of growth and decline of populations and resources; both are concerned with interactions between those populations and resources. Ecologists typically apply their concepts to the study of all organisms except humans; economists apply theirs to people, commodities, and money. Historically, there has been little overlap between the two disciplines. Recently, though, it has become more widely accepted that advocates of environmental protection need to consider economic development, while economists need to consider the environment. Both ecologists and economists have come to realize the mutual benefits to be had from a cross-fertilization of ideas. In this final chapter, we will examine the way in which conservation biology enters the domain of economics and policy, as we look at the first two of biodiversity's features listed above—its potential as source of jobs and income.

Many people have hoped that the sustainable exploitation of wild species that have economic value would provide incentives to conserve the habitat in which the species lived. Jan Pronk (the former Minister for Foreign Affairs in the Netherlands) and Mahbubul Haq (former Finance Minister of Pakistan), who drafted the Haque report on Sustainable Development that formed the basis for discussion at the Earth summit in Rio, point out that sustainable development is not simply a call for environmental protection, but a new form of economic growth that incorporates environmentally safe technologies and takes into account the scarcity of environmental resources. We have seen in Chapter 5 that attempts to use biological resources in a sustainable fashion are usually doomed to economic failure—most biological populations grow so slowly that financial benefits are most readily maximized by overexploiting the populations and investing the profit in alternative markets. The exploited population declines in number, perhaps to extinction. Yet the only alternative to sustainable use is unsustainable use, and that could bring about even more rapid declines. It seems we need to think again about how sustainable use can be made to work.

When a species such as a Tonga bean or a mahogany tree becomes a commodity, it acquires a monetary value. It can play a part in economic arguments about the most efficient or profitable way to make use of a natural habitat and its resources. But when evaluating such arguments, we must be certain that they consider the full value of a species, an ecosystem, or even all of biodiversity; otherwise, we cannot be sure that what we gain by exploiting a species or a habitat will be equal to what we sacrifice.

Putting a Value on Biodiversity

Attempts to quantify the value of natural resources are at the center of many economic, aesthetic, and philisophical arguments for the conservation of biodiversity.

As an indication of how poorly we value natural resources, let us look at the economic return from a hectare of converted rainforest. Chris Uhl and Geoffrey Parker have estimated that a hectare of rainforest supports about 800,000 kilograms (wet weight) of plants and animals. If we destroyed the rainforest to produce a pasture in which we grazed cattle, we would need about 6 square meters to produce the meat necessary for one quarter-pound hamburger. These 6 square meters could easily have supported one vigorous 60-foot tree (weighing about 450 kilograms) as well as 50 saplings and seedlings from 20 to 30 plant species, at least one perhaps containing an undiscovered new medical compound. All this vegetation would in turn support thousands of insects, from over a hundred species, and dozens of birds, mammal, and amphibian species would visit the patch of forest regularly.

There are three different ways we could assign a value to a single species. We could say that a species has a commodity value because it can be made into a product that can be bought or sold. Thus Brazil nuts have a value because their sale can supply needed foreign currency to Brazil and Peru, the countries in which they grow. Second, a species can have amenity value if its existence improves our lives in some way. We can experience delight in seeing a bald eagle, for example, or a rare species of orchid. Indeed, whole ecosystems can have both amenity value and commodity value because visitors enjoy the amenities but pay out money that benefits the local economy.

Finally, a species can be thought of as having moral value. This type of value is much harder to define, although some philosophers might say that all species have equal moral value in their own right. Perhaps the best definition is one we can take from Henry Thoreau, who believed that his ob-

▲ The mangetti nut is a staple food of the !Kung bushmen of the northwestern Kalahari, and an example of a potentially valuable forest product. Each year the bushmen return to the great white dunes on which mangetti trees grow to gather the protein-rich nuts.

The Financial Size of Several Industries		
Industry	Estimated world market	Growth rate
Forestry		
Tropical	7×10^9	
Total	85×10^9	1–2%
Pharmaceuticals	$120–200 \times 10^9$	2–5%
Tourism[a]	2000×10^9 (1987)	4–5%

[a] Tourism is estimated to employ 6.3 percent of the global work force.

servation of other species helped him to live a better life. Certainly many ecologists and evolutionary biologists would, I think, find themselves at least humbled by their own studies of the complexities of other species. However, quantifying moral values is a difficult problem for ecologists and economists alike. We can attempt to quantify the moral value that people attach to species by using a method that economists call contingent evaluation. In this process, they ask people how much they would be willing to pay to protect species quite independent of any use of that species. Surveys conducted using this technique indicate that species have not only a considerable moral value but one that can be measured in dollars.

Unfortunately, surveys of this sort indicate that the amount people are prepared to pay to conserve a specific species declines as we move from mammals and birds to the plants, insects, and worms that determine ecosystem function.

Although many species have considerable commodity, amenity, and moral value, the relative values of these are distributed unevenly among species, and indeed the assignment of such values has been attempted for only a handful of them. Given the overwhelming numbers in nature, and the fact that more than 50 percent of all species remain as yet undiscovered, it is unlikely to be an exercise that is applied to all species.

An alternative approach is to examine the current value of the total world market for commodities and services provided by tropical rainforests and other components of biodiversity. Forestry, fisheries, mining for mineral resources, harvesting of pharmacological plants, and tourism have all been explored as ways of using tropical rainforests. Each of these industries produces revenues of many billions of dollars a year on a global scale, and could be counted on to produce at least a few billion a year from the rainforest. The important question is, can these industries be implemented to give a sustainable yield, so that the forest is preserved?

Tropical Rainforests—Overexploit or Manage for Sustained Yield?

"Sustainable development" is the vogue term in international development at present, yet all too often what is meant by the term is not the sustained ecological yield that would maintain stock of an exploited resource at a constant level from generation to generation, but a sustained economic yield that will provide a constant financial return (one growing at the same rate as inflation). Obviously if the price of a commodity goes up as it becomes rarer, it is quite possible to drive a species to extinction while obtaining a constant economic yield from fewer and fewer individuals. Still, the advocates of this form of "sustainable development" view themselves as champions of conservation because their plans do not allow the uninhibited exploitation of a resource that would initially create an increasing, rather than constant, economic return. Unfortunately, their short-term calculations only consider profits and losses over the next five to ten years and thus cannot take into account the potential impact of losses of biodiversity on climate and the environment. While ideological debates between the advocates of sustained development and advocates of unregulated exploitation of natural resources smolder in the political arena, we are failing to carry out maintenance on the storehouses of environmental capital, and in many parts of the world we are actually subsidizing its destruction.

▶ Penan tribespeople on the island of Borneo pursue a traditional way of life that is endangered by the cutting of their home forest.

If we examine possible sustained uses of tropical rainforests and other natural resources, we shall see that although there are some examples of sustainable development that help maintain biodiversity, many fall short in this respect. The tropical timber trade is of particular importance to biodiversity because of the role it plays in rainforest destruction. Ironically, the total value of tropical timber exports, around $6 billion in 1991, is beginning to fall as commercially valuable forests are logged in order to meet demand from developed regions such as Japan, North America, and Europe. The timber trade has the most impact in those countries that gear production toward the export market. Just six countries accounted for 95 percent of the world's legal export of tropical timber products in 1988. Malaysia alone accounted for 62 percent of the exports and Indonesia for 22 percent. Singapore, the Philippines, Cameroon, and Liberia combined made up 12 percent. The $1.5 billion worth of timber exported from Malaysia, which supplies 90 percent of Japan's tropical imports, is having a dramatic environmental cost on the east Malaysian states of Sarawak and Sabah, on the island of Borneo. The International Tropical Timber Organization estimates that Sabah will be logged out by 1995, Sarawak by 2002, and Kalimantan, in the south of Borneo, by 2010. In Sarawak, continued logging will completely obliterate the homeland of the local Penan people, who are trying to halt the cutting of their home forests before it is too late. Unfortunately, the imposition of regulations in one country just leads the trade to move to other countries. Shortly after Thailand imposed a ban on logging in 1989, the Thai logging industry turned to neighboring Laos and Myanmar.

Alain Durning of the WorldWatch Institute suggests that to halt global deforestation, and reduce the rates at which other natural habitats are destroyed, three features of the economy need restructuring: (1) property rights to the forests, (2) the pricing of forest products, and (3) the decision-making power over forests. Although many endemic tribes live in tropical forests, they don't actually "own" those forests in any legal sense. They therefore have no say over the future of those forests, and indeed someone else with sufficient funds can usually buy the rights to use the forest, often by bribing government officials. The people who have always lived in the forest have more incentive to care about its long-term health, although they may not have the financial resources to maintain it in its current condition. In contrast, people who buy the forest as an investment usually want to see a rapid return on that investment. Similarly, the prices now paid for wood and other forest products do not reflect the ecological damage inflicted by timber extraction—soil erosion, decimated fisheries in areas downstream, and other environmental costs of production do not show up in the market forces. The result is that timber is underpriced and forests are undervalued. Finally, the power to decide the fate of a forest is typically concentrated not in the hands of the many people who depend on the forest for a livelihood, but in the few who profit from its deforestation and can sway political opinion accordingly. In particular, the rights of most traditional people are ignored. The pygmies of central Africa, who are probably the most ancient of forest dwellers, hold no enforceable claims to the forests they have inhabited for over 40,000 years. Similar inequities are dispossessing endemic tribes of the Amazon, where it is estimated that at least one major tribe goes extinct each year. The foresters who earn their livelihood in the boreal and old growth forests of North America and Europe would understandably like to have employment for the rest of their lives—the owners and managers who set high cutting quotas in these forests are far more responsible for the anticipated unemployment of these workers than spotted owls or other endangered species. The owners then use the substantial profits generated by overexploitation to pay for TV advertising and political campaigns that blame endangered species in future job losses.

An alternative way to use forests, which might preserve them while simultaneously providing a livelihood for those who live in them, is to extract nontimber forest products such as spices, rubber, nuts, and rattan. The revenues that these products generate can be substantial. Southeast Asia's rattan trade is worth more than $3 billion a year, nearly 50 percent of the total value of the world tropical timber trade. Exports of palm hats and Brazil nuts produced $20 million for Brazil in 1989. While there is no guarantee that alternative forest products will be harvested in a sustainable fashion, it seems likely that they will be extracted in ways that are less damaging then the clear cutting that is taking place at present.

How does the of use of nonwood or "minor" forest products compare economically with the use of timber resources? A number of studies have attempted to compare the net profit from harvesting an area of rainforest as timber with the longer-term yield obtained by harvesting forest products such as fruits and fiber as they appear.

One group of ecologists have undertaken a survey of the economic value of several one-hectare plots of Amazonian rainforest in the Mishana district near the city of Iquitos in Peru. The inhabitants of the Mishana region are detribalized indigenous people who make their living practicing shifting cultivation, fishing, and collecting forest products to sell. The survey found that on average one hectare of forest contains 842 trees (plants greater than 10 centimeters in diameter), representing 275 species from 50 different families. Of these trees, 350 individuals (42 percent) from 72 species (26 percent) yielded products with a market value in the nearby town of Iquitos. Seven tree species and four palms produced edible products, while 60 species had timber value, one species produced rubber, and many lianas, palms, and smaller plants were of medicinal value.

The ecologists calculated that one hectare of forest produces fruit worth $650 each year, while the annual yield of rubber amounts to around $50. If we take into account the costs of harvesting and transporting the yield to market, then the net annual revenues from fruit and latex are $400 and $22 per hectare. The forest also contained 93.8 cubic meters per hectare of harvestable timber worth around $1000 at the sawmill or, if harvested on a 20-year cycle, worth about $310 each cycle, for a profit of $490 (assuming a 5 percent inflation rate). Compare this figure with the net value of the annual fruit and rubber harvest over the same twenty-year period, a total figure of $6330 (also assuming a 5-percent inflation rate). This calculation implies that the continuous harvests of fruits and latex represent more than 90 percent of the forest's value, while the timber harvest is a marginal, one-time crop. Although it is possible to cavil with some of the details of this calculation, it raises the hope that the economic value of the forest, once fully appreciated, can secure its protection rather than encourage its destruction.

We have seen that most exploited natural systems are susceptible to capitalization—the tendency toward overexploitation that results when the growth rate of the managed resource is less than the growth rate of capital if invested in alternative ventures. Many cases of overexploitation occur when misguided aid schemes attempt to develop industrial complexes that utilize forest or mineral resources whose economic goals are set by the financial standards of their home markets. The economics of sustained rainforest use will depend not only on the intrinsic growth rate of the species you wish to exploit and its interaction with other species, but also upon the type of alternative investments you have access to and the general economic health of the country in which you live. The tendency to overexploit resources may be considerably higher in countries with high interest rates and low human life expectancy than it is in countries with more stable economies. In contrast, there may be less tendency to overexploit resources that can only be bartered locally. Indeed, in some places the forest may act as a "bank" for the local inhabitants; here the absence of alternative sources of investment may considerably reduce the propensity to capitalize and overexploit the available resources.

A Dyak community living in West Kalimantan on the island of Borneo has been described by Christine Padoch of the Institute of Economic Botany, at the New York Botanic Garden. The community has demonstrated that sustainable use is easily possible in rainforests, providing its inhabitants take a long-term economic view. The Dyaks have considerably increased the rainforest's natural productivity in large areas surrounding their longhouses. These areas are transformed over many years and several human generations into plant communities that may more accurately be thought of as gardens. The Dyaks plant fruit trees around their houses, and

▶ Dyak women carry durian fruits harvested in their forest gardens. The Dyaks take thousands of these spiny fruits, which are highly sought after for their aromatic, creamy pulps, to market during the harvest season.

in forest clearings around abandoned dwellings. Particular tree species are occasionally transplanted from distant locations, or to initiate cultivation the Dyaks may simply spit out the seeds of a fruit they're currently eating. The Dyaks clear away vines and other potentially competing species that are not a source of food. They plant as many as 74 different species of fruit tree, and recognize more than 100 edible species. The location of particularly prized individual trees are known to several generations of the tribe. In some cases, when a tree with highly prized fruit such as the durian *(Durio zibethinus)* grows up and shades a house, it can prevent the owners from drying their rice crop, which would then rot in the humid climate. Rather than chopping back the tree, the house is moved to another location. Felling a durian tree is regarded as crime in Dyak society.

Unfortunately, government policies regard the ways of the Dyaks as underdeveloped, even though their careful habits allow the density of the human population in the forest to exceed that in other tropical forests by a factor of more than 80. The government is offering grants to encourage rubber plantations and other monocultures that will raise small amounts of foreign currency, but increase the dependence of the region's inhabitants on imported foods. Luckily sections of the Dyak community have successfully resisted this change and are making substantial profits from the growing international market for durian fruits.

Medicinal Plants: Does Money Grow on Trees?

Many plants and animals of the rainforest contain chemicals that are or could be used as anesthetics, stimulants, or other drugs. At present, around 70 percent of the drugs used in modern medicine are modeled on natural compounds; yet these are derived from only around 250 plant species. In fact, fewer than 0.1 percent of plant species have been examined for their medicinal properties. Chemical prospecting for plants and animals theoretically offers huge potential as a way of both using and conserving biodiversity, because the sale of drug-producing species could raise funds to help conserve the habitats in which species live.

Tourists and visiting ecologists can easily imagine the mysterious, complex, green world of the tropical rainforest as a place in which plants grow with untold magically medicinal powers. The indigenous people who live in the forests do not have to imagine; their traditional healers have always gathered plants for use in medicinal treatment. The World Health Organization estimates that 80 percent of the world's population relies on traditional plant-based medicines. It is the hope of many people from developed countries that we can learn from the shamans who garner this knowledge—and that the value of preserving rainforests will be demonstrated beyond question. We have already learned to use rainforest plants to treat many devastating diseases, including childhood leukemia, Hodgkins disease, glaucoma, and heart disease.

Unfortunately, a mere 1100 of the 365,000 known species of plants have been examined for medicinal properties. Some way is needed to increase the rate at which species with medicinal properties are identified, and three different strategies have been suggested: random search, further examination of species from families with known biotic action, and interviews with indigenous healers. The last of these has proved the most efficient; in a comparative study, species collected at random were four times less likely to show activity against the HIV virus (the virus causing AIDS) than species classified as "very powerful plants" by Don Aliho, a traditional herbalist in the Brazilian Amazon. Although these results are encouraging, several factors impede the race to conserve this knowledge. Don Aliho and many other healers are in their 80s and 90s. Although they have many grandchildren, few of them have had apprentices, and little of their knowledge, acquired over many years, is written down. When they die, it is likely that this wealth of information will be lost. Furthermore, as rainforests shrink, traditional healers have more difficulty locating the plants they need. A midwife specialist in Brazil suggested that the time she needed to locate and retrieve any specific plant had doubled in the last five years and increased by a factor of seven since her apprenticeship in 1940.

► A Mayan herb collector in Belize gathers materials that will later be used by shamans to create medicinal treatments.

Just how big is the potential for tropical (and temperate) forests to supply new medicinal drugs? So far, one out of every 125 plant species studied has produced a major drug. The market value of these drugs in the United States is at least $200 million per year. If we estimate that we lost one tree species a day, then we lose three to four potentially valuable new drugs every year, at a total cost of around $600 million. Yet most drug companies ignore the forest as a source of drugs. Instead, they prefer testing randomly synthesized chemicals, even though the chance of finding a major new drug is on the order of one in 10,000 for each compound tested. The companies have a much harder time patenting natural products than ones they have synthesized, so their potential profits are diminished when they attempt to market natural compounds. Their aversion to natural drugs is also due in part to the litigation that could follow the potential misuse of drugs in the United States, but also to the trade barriers and tariffs that the pharmaceutical industry itself has lobbied for to prevent the import of drugs based on plant products.

Despite the significant trade and legislative barriers, the overwhelming majority of prescription drugs used at present in the United States are based on natural products. In a fascinating review of the country's top prescription drugs, Francesca Grifo of the American Museum of Natural History has shown that 118 out of the top 150 prescription drugs were originally derived from living organisms: 74 percent from plants, 18 percent from fungi, 5 percent from bacteria, and 3 percent from vertebrates (indeed, all drugs derived from vertebrates are from one species of poisonous snake, *Bothrops!*). Of the top 10 prescription drugs in the United States, 9 are based on natural plant products. The figures are likely to be significantly higher for the use of natural products as drugs in other parts of the world, since the United States makes available a particularly undiverse selection of pharmaceuticals.

Many drugs that are widely used in Europe are not even for sale in the United States. For example, a plant derivative based on the leaves of the Ginkgo tree is now used by 80 percent of Europeans over the age of 45. The species escaped extinction in the wild when preserved in monastery gardens where they were tended for many centuries by Chinese monks. Compounds distilled from leaves of the ginkgo tree increase the rates of cerebral blood flow and are believed to help prevent senile dementia. Another drug for sale in Europe, but not in the United States, is an extract of mistletoe that has been shown by laboratory tests to double the survival time of women suffering from breast cancer.

A number of recent studies have highlighted the potential dangers posed to human health by the continued assault on the environment, particularly the destruction of tropical forests. In particular, new pathogens such as HIV and the Ebola virus are appearing at a time when we are destroying a major resource that could supply new drugs with which to treat these diseases. The emergence of new diseases, and the stubbornness of old ones, will keep the demand for new drugs at a high level. Most people in the tropical countries where most of the world's biodiversity lies still suffer from a large variety of parasites and other infectious disease. Many millions of children continue to suffer from parasitic intestinal worms such as hookworm, for example. In countries where parasites are also chronic in the adult population, the strength of adults is sapped to such an extent that industrial productivity and economic growth are retarded. Few things will convince people more of the value of biodiversity than if it can be shown to have a significant impact on human health and well-being. In particular, drugs to assist with the world's biggest problem may lurk in the rainforest: a number of tribes endemic to rainforests in different parts of the world, such as the Kai'por Indians in Brazil, have discovered plants that may be used for human birth control. In some cases, the ingestion of only a single leaf ensures that a woman will never become pregnant again!

Movies such as *The Andromeda Strain* and *Outbreak* as well as books such as *The Hot Zone* have made the public aware that new pathogens continue to emerge. The growing human population is probably the main factor determining where and when such pathogens appear and become established in humans. Essentially, the more human beings that are out there sampling the environment, the greater the chance that some of them will pick up a new pathogen and transport it to a populous area where it can cause an epidemic. However, it is likely that parasites and pathogens that have been with us for a long time will remain the greatest threat to the

▲ The leaves of the ginkgo tree supply compounds taken by millions of people to prevent senile dementia.

welfare of human populations—particularly if a changing climate allows pathogens to become established in new areas that have become noticeably warmer and damper. The search for new compounds will become increasingly frantic as these pests and their vectors evolve resistance to many of the manufactured compounds used to control them. Resistance will develop with particular rapidity to laboratory-synthesized drugs that, as is often the case with such drugs, are mere variations of compounds, similar in chemical structure, to which resistance has already evolved. In contrast, drugs derived from natural products are likely to have novel properties to which the pathogens have not developed much resistance. Furthermore, as many natural compounds consist not of one but of a large number of active ingredients, using such a compound will be like using more than one drug to treat an outbreak—if more than one of the chemical constituents has active properties—and will therefore considerably reduce the rate at which resistance evolves.

The chemical company, Merck & Company, has recently signed an agreement with the government of Costa Rica to encourage biological prospecting for drugs in the forests of that country. At the root of the IIVBio agreement is the hope that biodiversity can be saved by discovering products that will illustrate its considerable value to humans as a source of drugs. The agreement is not perfect in that it assumes the government rather than the endemic people are the natural owners and protectors of the lands, but it remains an important first step.

Although the promise of medical cures is an effective argument to rally public support for rainforest protection, there are many difficult and fascinating problems involved. One of the biggest is that there is presently no international legal means for the tropical countries to receive concrete benefits from the development of a plant into a pharmaceutical product accessible to westerners who can afford it. Of course, a price could be charged for all plants removed for the manufacture of a drug, but drug companies rarely use the actual plants in manufacturing. It is often more profitable for a drug company to synthesize copies or simplified analogies of any active drug isolated from a plant or animal. The company can then produce large quantities of the substance in industrial laboratories and forget the worries such as variation in quality associated with exploiting a natural source of the compound.

One solution would be to grant intellectual property rights to the indigenous people who may have first "discovered" the plant's medicinal value, so that they would receive a royalty on its sale. Unfortunately, not a single country in the world extends intellectual property rights to indigenous ecological knowledge. As Alain Durning of the WorldWatch Institute has pointed out, if a traditional healer knows how to cure a skin disease with an herbal remedy, that is called folklore. If a pharmaceutical company isolates and markets the active chemical in the healer's herbs, it is called a medical breakthrough and is protected by a patent and rewarded with international monopoly power. Biological prospecting will only be successful as a conservation measure when the drug companies follow the practices of the handful of ethnobotanists who study the plant uses of traditional healers. These ethnobotanists follow codes of conduct that include negotiating with local communities to ensure that they benefit from any commercial products that their knowledge help to create.

The International Convention on Biological Diversity of 1992 has laid down important guidelines for biodiversity prospecting to ensure that a significant proportion of the financial benefit goes back into protecting the habitat and the indigenous tribes that are the source of this potential wealth. Although more than 106 countries have ratified this convention, the United States has not. Much of President Bush's reluctance to sign the biodiversity treaty was a response to pressure from drug companies that felt their "competitiveness" might be threatened by a treaty that acknowledged the potential utility of plant-based drugs with a tropical origin.

How do we marry incentives for a pharmaceutical company to spend the money required in "screening" for medicinally valuable plant products with incentives for the host country to invite the company to conduct the search? These incentives are seemingly at odds with each other. The pharmaceutical company requires assurance of a profit, especially given the risk of failure and the long stretch of time—often 5 to 15 years—needed to bring a successful plant medication into the market. The host country, on the other hand, also wants to profit, especially since enormous amounts of plant material may be required to produce useful amounts of medicine and since certain plants (e.g., the rosy periwinkle) may yield billions of dollars of profits. It will only be possible to keep both sides happy if development costs are kept low, so that a drug company can get a natural compound out of the forest cheaply and quickly enough that it can afford to share its profits.

Early attempts to locate useful plants were often unsuccessful. Researchers would accumulate bundles of moldy plants, some of which possessed interesting properties, but by the time these were identified it usually never proved possible to find the plants again. Lisa Conte, the founding managing director of Shaman Pharmaceuticals (San Carlos, California), has strongly emphasized the importance of arranging reciprocal exchanges of western medicine for ethnobotanical knowledge. Conte suggests that to keep development costs low, residents be recruited to extract plants near where they live. It may be sensible to concentrate on examining weedy species, as these will be common in areas close to human disturbance. The chances of locating economically valuable compounds are much higher if searches are restricted not to miracle cures for AIDS or cancer, but to drugs for ailments such as fungal infections, cold sores, and herpes. These infections are common in both developed countries and the rainforest, and there is a large market for drugs used to control them.

Ecotourism

Tourism is probably the world's largest industry and is potentially a very powerful way to make biodiversity pay. In at least five countries (Kenya, Ecuador, Costa Rica, Madagascar, and Nepal), it is the major earner of foreign currency. The gorillas in Rwanda were a major tourist attraction before the recent civil war; their value as a future economic commodity has been fairly effectively maintained throughout the war, as both sides would like to capitalize on ecotourism once civil strife has declined.

The United States exports around 5 million "ecotourists" each year, each of whom spends 2 or 3 thousand dollars, and their numbers are expected to increase by at least 20 percent in the next five years. Studies in Kenya indicate that land in national parks such as Amboseli raises $40 per hectare each year, as opposed to the $0.80 it would raise if used for agricultural purposes. Coarse calculations suggest that each lion in the park raises about $20,000 per year and that the elephant herds annually raise about $610,000—considerably more than their one-time value as a source of ivory. Unfortunately, no similar figures are available for rainforest beetles,

▶ Tourists in Zaire, gazing in fascination at a gorilla, are probably too close for the good of either the animal or themselves.

but it should soon be possible to make similar calculations for national parks and their natural residents in other countries.

There are significant asymmetries in the way that the costs and benefits of ecotourism are distributed. Like all big businesses, ecotourism tends to focus profits to a few while diffusing the costs to be paid over many. Most of the financial benefits accrue to the tour companies that organize the trips. Unfortunately, these are usually not based in the country the tourists have chosen to visit. Similarly, there are huge asymmetries in the impact that tourists have on different species. Studies at Amboseli National Park in Kenya indicate that 80 percent of vehicles in the park use only 10 percent of the roads and that each tourist spends around 45 percent of his or her time looking at lions and cheetahs (predators that are usually asleep!), 12 percent at elephants, and 7 percent at rhinos. The average visitor to the park spends less than half a minute looking at any of the other species in the park.

A system of ecotourism that achieves sustainable development will meet a number of criteria. First, tourism should create opportunities for local people. In some places this is already the case: in the Peruvian Amazon, for example, about 60 percent of the $1.3 million brought in by tourists goes toward local salaries and the purchase of supplies on the local market. Second, tourism should also help justify conservation to local governments. In Nepal, a project to build a hydroelectric dam near the Chitwan nature reserve was canceled after an economic analysis showed that the park raised more dollars than would the hydroelectric scheme. Third, tourism should provide funds for purchase and maintenance of reserves. This was clearly the case in the forests of Rwanda before the recent civil war; the gorillas raised around a million dollars a year in tourist fees, while the park guards cost only $150,000 per annum. Similarly, revenues from the Galapagos Islands pay the park budget for all of Ecuador. However, although the government attempts to restrict tourist numbers to 25 thousand per year, as many as 50 to 70 thousand people a year actually visit the islands. The extra thousands of people are eroding the paths, disturbing native species, and reducing visitors' wilderness experience.

Yet even tourism may have detrimental effects. Every park has a carrying capacity for tourists; as their numbers increase the fees they pay are often too small to cover the damage they do. The littering, trampling of foot paths, and overuse of roads can be repaired given sufficient revenues, and can actually create local opportunities for employment. However, the indirect damage tourists cause by disturbing the animals and preventing their access to resources is less easily remedied. This damage is harder to quantify, but the loss of mating and feeding opportunities caused by the presence of a natural enemy may have as significant an impact on a species as its actual exploitation. These influences can be quite subtle—studies by Malcolm Hunter of red pandas in Nepal revealed that these wonderfully charismatic animals were a major draw for tourists who had come trekking in the Himalayan foothills, even though their chance of seeing one was small. One by-product of the increase in tourism is a demand for cheese and other dairy products—most tourists have discovered there is only so much yak butter you can eat and still enjoy a hiking vacation. To satisfy the demand for dairy products familiar to the tourists, deforestation has increased to create pastures for cattle. The loss of high alpine forests has begun to reduce red panda numbers and has further lowered the tourists' chances of seeing one.

Many national parks undervalue their resources and charge entry fees that are too low to cover the cost of park management. For example, studies by Dominic Moran of University College in London reveal that Kenya earns about $400 million a year from tourism—about 40 percent of the country's foreign exchange earnings in any one year. Yet Kenya Wildlife Service, relying on money collected from $20 park fees, receives only $13 million a year. Although most tourists spend several thousand dollars on their vacations, less than 1 percent of this money finds its way into protecting biodiversity. In surveys, tourists say they would be happy to pay substantially higher fees to enter the national parks. Increasing the fees to tourists, while maintaining lower fees for local residents, could considerably increase park revenue, while also reducing the damage that can occur when an excess of tourists visit a park.

A less treatable problem in many countries is the insatiable demand of tourists for water to drink, wash, and swim in. In arid countries where water is in short supply, the demand can monopolize a large proportion of the available resources. The most recent figures published by the United Nations Fish and Agriculture Organization estimate that 15,000 cubic meters of water are needed to irrigate one hectare of high-yield rice; alternatively, this volume of water is enough to supply 100 nomads and 450 cattle for three years, or 100 urban families for two years. Simply add tourists to watch the water evaporate—100 guests at a luxury hotel will use this much water in 55 days!

A final criticism aimed at ecotourism is that tourists are capricious in their choice of location. Some will go only where few have gone before, others will consistently go to a few well-tried places, while many others will visit only places where they can meet other tourists. Their inconsistency creates the potential for boom-and-bust cycles in the local tourist economy, particularly when tourists feel that local politics may put them in some jeopardy. The development of ecotourism itself may go a long way toward rectifying this final problem. The delicate and potentially dangerous political situations that worry tourists are usually a consequence of huge inequalities of income between the few wealthy individuals and the many very poor ones—inequalities that are often a feature of economies that depend upon extracting natural resources. The considerable numbers of jobs contributed by tourism can help reduce the inequality of income. More jobs and less inequality of income could potentially make countries more stable and hence more attractive to tourists. Stable developing countries may in turn attract investment in education, light industry, and engineering that can further build their economies.

Debt-for-Nature Swaps

Some rainforests have proven themselves valuable in an unexpectedly indirect way—as a form of payment for debt. An important new way of buying land to be set aside as nature reserves is the "debt-for-nature" swap; an idea originally developed by Thomas Lovejoy of the World Wildlife Fund, now at the Smithsonian Institution. Debt-for-nature swaps are designed to relax debt's stranglehold on development and reverse the unsustainable use of tropical forests. They were inspired by the insight that many tropical countries are deeply in debt to nations who have supplied arms and other luxuries of development. Brazil, for example, had $111 billion in foreign loans in 1986 and also holds 30 percent of the world's tropical rainforest. Many of these countries see little choice but to destroy some of the world's most important natural assets, particularly tropical rainforests, in order to raise funds to meet these debts. As Thomas Lovejoy has put it, "Under the best circumstances, debtor nations find it hard to address critical conservation issues because of multiple social needs. Stimulating conservation while ameliorating debt would encourage progress on both fronts."

There is a fairly distinct relationship between rates of rainforest destruction and levels of debt. Developing nations owe the phenomenal sum of $1.2 trillion, money that is difficult for these nations to raise. Mexico reneged on its debt in 1982 and other countries attempted to follow suit, raising doubts about whether the debt would ever be collected. Banks have attempted to recoup at least some of the losses by selling debt at a discount. In a debt-for-nature swap, a conservation organization acquires the debt at a discount and asks for the debtor country to redeem the debt by supplying land for reserves and salaries for people to manage, monitor, and protect those reserves.

In July 1987, Bolivia entered the first debt-for-nature swap with Conservation International. Conservation International purchased $650,000 of Bolivia's debt for $100,000 (15 cents on the dollar). In exchange for canceling the debt, the president of Bolivia agreed to demarcate 3.7 million acres of tropical forest as an extension of the existing Beni Biosphere Reserve and establish a $250,000 fund in local currency to manage the biosphere reserve. In December 1987, the World Wildlife Fund concluded an even larger debt-for-nature swap with Ecuador. Such swaps have now been negotiated with other countries, including Costa Rica and Madagascar.

These projects must be structured flexibly to meet the financial and ecological interests of both sides. In particular, the large volumes of money swapped do not help local anti-inflation policies. Furthermore, many countries feel they are being denationalized by the process.

Ultimately, the environmental destruction driven by the demands of burgeoning trade cannot be fully halted without major debt relief for developing countries. Allowing debt-strapped third-world producers to escape this debilitating treadmill would make it possible for them to encourage democracy and turn to the kinds of small-scale grassroots development efforts that are capable of reducing both poverty and environmental degradation. These small-scale projects are also more likely to achieve sustainable use.

Reforming World Trade

Hilary French of the WorldWatch Institute has recently examined the relationship between trade and the environment. More than a quarter of the world's trade involves goods derived directly from natural resources such as timber, fish, and minerals, called primary products. Most developing countries are net exporters of food, raw materials, minerals, and fuels to the industrial world. Primary products constitute more than 98 percent of the total exports of Bolivia, Ethiopia, Guyana, and Nigeria, compared with 24 percent of United States exports and only 2 percent of Japan's. Developing countries are thus particularly vulnerable to aggressive trade practices that play one country against another as sources of raw materials—in many cases countries are forced to lower their prices to ensure they obtain contracts; they must then increase the damage inflicted on the environment in order to minimize the cost of extraction.

With such downward pressures on the prices paid for primary resources, it is no wonder that there are huge inequalities in the economic returns from these resources. In particular, timber and minerals are often exported at prices that underrate their economic value, and much of the profit on their final sale is accrued in the developed country importing and processing the product. There are, moreover, significant discrepancies in where the costs of their extractions are paid. A study by the International Institute for Applied Systems Analysis in Austria found that mining bauxite, ore that is ultimately made into aluminum, uses about 10 percent of the total energy expended in aluminum production and produces 90 percent of the wastes, while accounting for only 10 percent of the profits. The second stage of processing, which turns alumina into raw aluminum and is also usually undertaken in the developing country, uses 80 percent of the energy and produces 9 percent of the waste while accounting for 20 percent of the profits. In contrast, the final stage of manufacturing, usually performed in a developed country, generates 70 percent of the profits while requiring only 10 percent of the total energy and producing only 1 percent of the waste. Developed countries are able to thus maximize their profits and minimize the environmental damage to their own countries.

One way that developed countries manage to keep hold of the profitable final stages of manufacture is by imposing economic tariffs on the importation of processed materials. It is hoped that this protectionist practice may be rectified in the present GATT convention, since developing countries are also hindered from exporting processed forms of cocoa, fish, minerals, and rubber. If these barriers were removed, these countries could make a more significant profit from the sustainable use of their natural products. If such tariffs continue, they will have little choice but to destroy their natural resources and pollute their environments in a way that is only just economically viable.

These examples suggest that the official statistics that describe a country's trade, and are used as economic indicators, should include the amount of damage this trade is doing to the environment, as an index of the country's longer-term well-being. Although the annual growth rate of Indonesia's gross national product between 1971 and 1984 was 7.1 percent, once the deterioration of its forest, soil, and petroleum assets were figured into the calculation, the growth rate declined substantially to 4 percent.

Trade has always played a central role in the development of human civilization, from the spread of Islam into Africa, the rise of the European city-states, and European expansion into the Americas, to the emergence of Japan as a new economic superpower. As we enter the twenty-first century, with the environment in crisis, it is crucial that trade practices be reconciled with the needs of the environment if both the world economy, and perhaps the human species, are to continue to develop healthfully. Changing the way the world's economy works is plainly not a simple problem, nor is anyone who proposes or supports such a radical notion likely to be popular with the people who are making a significant profit in the current economic order. Economic change that recognizes the importance of biodiversity and the services it provides is likely therefore to be slow.

One sign that more people are taking seriously the need to preserve biodiversity has been the rising membership of voluntary organizations dedicated to preserving wilderness and wildlife. Membership in the Wilderness Society has shot from 40,000 at the start of the 1980s to 320,000; Friends of the Earth has seen its members double in each of the last few years. The WorldWatch Institute has now over a thousand applicants for each job. Furthermore, both the European community's Maastricht Treaty and the North American Free Trade Agreement (NAFTA) recognize that the pursuit of sustainable development and strengthened environmental policies are vital to the success of trade and development

▶ This poster advertises fines for capturing the endangered St. Lucia parrot. Thanks to a widespread education campaign in schools, every child on the island recognizes the national bird.

within these communities. These international treaties have begun to acknowledge that in fairness to later generations we should leave our children not only a balanced budget but also an intact environment and a stable world.

One powerful tool that governments can use to protect the environment is taxation. David Roodman of the WorldWatch Institute has shown that, in many countries, minimal taxation on natural resource use and environmental destruction actually encourages overexploitation of forests and fisheries, while also imposing minimal costs on those who pollute the environment. Roodman suggests that tax laws be adjusted over the next twenty to thirty years to place the costs of environmental exploitation and pollution on the corporations that profit from this exploitation. Employment taxes could be reduced for industries that employ people to clean up, recycle resources, and rejuvenate the environment. A final step would be to reduce taxes and death duties for property owners who encourage the maintenance of natural biodiversity on their land. In many developing countries, reversing the current tax subsidies that encourage environmental destruction could weaken the political power of the large resource extraction industries who use their profits to support the campaigns of helpful politicians, and it might even give democracy a more realistic face in many developed countries.

Ultimately, there are no simple solutions to the problem of how to achieve both economic development and sustained use of biodiversity. Almost all types of natural product seem prone to overexploitation. Moreover, extractive reserves still tend to reduce biodiversity, and if improving human welfare is the goal, they are no substitute for programs to redistribute existing farm land, reform farming practices, and halt human population growth. Although this chapter has suggested several ways of valuing the natural environment, each of these approaches still probably underestimates the long-term value of biodiversity. Bryan Norton of the University of South Florida has suggested that the net value of biodiversity is equal to the summed value of all gross national products of all countries from now until the end of the world, because human life and economies are in essence entirely dependent upon biodiversity. As Harold Morovitz of Yale University has pointed out, one answer to the question "How much is a species worth?" is "What kind of world do you want to live in?"

Protecting the world's remaining biodiversity will require scientists and policymakers to work together in innovative ways. The scientific problems involved will often be complex: indeed, the mathematics of ecosystem function and the calculus that governs the maintenance of biodiversity present as great a challenge to science as unraveling the genetic code, determining the structure of the atom, or fathoming the origin of the universe. Because scientific understanding of ecological problems will have an immediate impact on the quality of life on earth, it is sobering that we have left the development of such an understanding until so late. Ultimately our tardiness might turn out to be an understandable consequence of the difficulty of the task: some of the world's most challenging scientific problems are the ones studied by ecologists and conservation biologists.

Ultimately we can only conclude that other species have a value that cannot be measured in the simple terms used to describe the economic benefit that we derive from their exploitation. The earth's biodiversity and our own species are the products of the same evolutionary process, and to many people the presence of other species provides a tangible and satisfying sense of our own place in the world. The Biophilia hypothesis has been proposed by E. O. Wilson of Harvard University to describe our innate appreciation and love for nature and natural diversity. Wilson suggests that the human need for nature is comparable to religious awe or the regard we feel for great works of art or music. Each of us is appalled when we hear that a great painting of sculpture has been destroyed. We flock to see the last performance of a great dancer or musician, and we are always amazed at the number of pieces of great music and literature that have remained unrecognized during their creator's lifetime. Yet the humblest beetle species is still the product of a considerably longer creative process than any painting, statue, or piece of music. If the Biophilia hypothesis is correct, then the natural habitats being destroyed today and the species lost are crucial to our spiritual well-being, in addition to being vital to the earth's long-term health.

*E*pilogue

The Mayan ruins at Tikal in Guatamala are one of the best places to observe Central America's diverse wildlife. The trees planted by the Maya to supply fruit and nuts have grown over the wreckage of the city and provide an almost continuous supply of food for the birds, insects, and monkeys that are everywhere in great abundance. The ruined courtyards provide a spectacular site for the courtship display of the ocellated turkey, while the pyramidal temples provide useful vantage points for staring into the canopy layers of the tropical forest. The view at sunrise across the Yucatan peninsula from the top of the tallest temple is breathtaking; the haunting calls of howler monkeys sound like a mad hurricane chasing across the forest, and the screams of the orange-breasted falcons echo around the courtyard several hundred feet below. This apparently untamed jungle was once the center of a major human civilization that stretched throughout Central America. One thousand years ago, the view from the top of the temple at Tikal would have been analogous to the view we see today from the top of the World Trade Center in Manhattan. In both places, human beings displaced the natural habitat and created vast temples as symbols of their success and dominion.

Our consideration of the present-day scene from the summit of the ruins at Tikal should give us pause for thought. Why did the Mayan empire collapse? Is it possible that something similar could happen to the vast empires of trade that radiate around the world from Manhattan and other urban centers? What will the view from the top of the World Trade Center look like a thousand years from now?

In Tikal, the tropical forest has slowly reclaimed the elaborate courtyards and pyramids built to house the Maya and provide them with places of worship. No completely satisfactory explanation has yet been advanced to explain why the Mayan empire collapsed; recent evidence indicates that a prolonged drought, which could have been partially induced by the widespread deforestation around Tikal, might have played a role. The trees may have been burned to provide the heat needed to create the stucco used in the elaborately decorated veneer that covered the Mayan temples. It takes the wood from around 20 large trees to produce a small, meter-high pile of the limestone that forms the basis of this stucco. The Maya would have had to cut and burn vast tracts of forest to decorate their increasingly elaborate temples, and also to obtain wood for building houses and making fires. The deforestation would have led to soil erosion, which in turn would have reduced crop harvests, perhaps producing widespread hunger, civil unrest, and the series of civil wars that may in part have led to the demise of the Classic Mayan civilization around A.D. 800.

Tikal is by no means the only ancient civilization to collapse as a result of ecological degradation. A similar fate befell the population of Easter island, a potential paradise in the Pacific ocean. The small population that colonized the island in the fifth century had grown to around seven thousand people by 1550. During these centuries of growth, the population fissioned into several competing tribes that erected large ceremonial statues on different parts of the island. Sometime in the sixteenth century the population of Easter island collapsed dramatically. Archaeological evidence again suggests that the decline followed the removal of the island's forests to supply agricultural land for the increasing human population, and to provide tree trunks on which to roll the rock for the statues across the island. The complete deforestation again led to erosion and the loss of topsoil. With no wood left for building houses or constructing canoes, the population could no longer fish or escape to another island, and were forced to live in caves dug in the hillside. Eventually, there was no firewood left with which to cook food, and the society collapsed into anarchy with each tribe raiding and pulling over the statues of the others. When European explorers first visited Easter island in the 1830s, the few hundred people still living there had reverted to a primitive Stone Age culture. They could not even recall how their ancestors had moved the giant rocks from the single quarry that supplied their stone.

Most early human civilizations seemed to have collapsed into civil strife following some form of overexploitation of natural resources. The earliest settlers in the valleys of the Tigris and Euphrates ended by overexploiting the land to provide food for a steadily growing human population. The irrigation schemes essential to farming in the area eventually caused saltation of the soils along the river valleys, which in turn led to malnutrition, civil unrest, and disease. The once populous areas where primitive agriculture first developed were then slowly abandoned, and new areas of forest along the Nile and other parts of the Mediterranean were colonized and converted to agricultural land.

In the past, human populations have escaped local overexploitation of resources by migrating into new areas. There are now no new continents or islands to migrate to; we are as isolated as the natives of Easter island in the fifteenth century. If we wish to avoid their fate or that of the Maya, we have to find ways of maintaining the biodiversity that allows us to survive. A necessary goal to achieve this is to find ways of controlling our own population growth.

Today, Tikal, Manhattan, and Easter island provide three different perspectives on what can happen to the earth and human society. Nature has temporarily reclaimed Tikal so that visitors are able to appreciate its rich tropical forest. The decaying Mayan ruins remind us of the frailty of human civilization; its maintenance is much less solidly dependent upon might and innovation than it is upon the rational management of the complex diversity of other species that supply our essential needs. The alternative prospect is Easter island, which remains ecologically impoverished, its famous statues acting as monuments to human inability to recognize our dependence upon other species. If we continue to allow our population to grow and exploit the earth's natural resources at increasing rates, then it is hard to see how we can avoid a fate similar to that of the natives of Easter island. As an ecologist and evolutionary biologist, I have a residual optimism that if the human population collapses, then the surviving components of biodiversity will recolonize the planet, much as they have done at Tikal. Of course, none of the species we have extinguished will evolve again, and neither will our own. In a time shorter than the two to five million years for which humans have inhabited the earth, our only legacy will be a slowly recovering but biologically impoverished small blue planet. I hope that this book has provided a short, optimistic introduction to the science that provides an alternative solution to the pessimistic journey from Manhattan to Tikal and Easter island.

Bibliography

Ballou, J., and K. Ralls. 1982. Inbreeding and juvenile mortality in small populations of ungulates: A detailed analysis. *Biol. Conserv.* 24:239–272.

Beddington, J. R., and R. M. May. 1982. The harvesting of interacting species in a natural ecosystem. *Sci. Am.* 247:62–69.

Berger, J. 1990. Persistence of different-sized populations: An empirical assessment of rapid extinctions in bighorn sheep. *Con. Biol.* 4:91–98.

Bibby, C. J., et al. 1992a. *Putting Biodiversity on the Map: Priority Areas for Global Conservation.* International Council for Bird Preservation, Cambridge.

Bierregaard Jr., R. O., T. E. Lovejoy, V. Kapos, A. A. dos Santos, and R. W. Hutchings. 1992. The biological dynamics of tropical forest fragments. *Bioscience* 42:859–866.

Bradshaw, A. D. 1984. Land restoration: Now and in the future. *Proc. R. Soc. Lond. B.* 223:1–23.

Burgman, M., S. Ferson, and H. R. Akcakaya. 1993. *Risk Assessment in Conservation Biology.* Chapman & Hall, London.

Burney, D. A. 1993. Recent animal extinctions: Recipes for disaster. *Am. Sci.* 81:530–541.

Cairncross, F. 1989. Costing the Earth: A survey of the environment. *Economist* September 2, 1–19.

Caughley, G. 1994. Directions in conservation biology. *J. Anim. Ecol.* 63:215–244.

Caughley, G., and A. Gunn. 1995. *Conservation Biology in Theory and Practice.* Blackwell Science, Oxford.

Cherfas, J. 1984. *Zoo 2000.* BBC, London.

———. 1988. *The Hunting of the Whale.* Penguin, London.

Clark, C. W. 1973. The economics of overexploitation. *Science* 181:630–633.

Cohen, J. E. 1995. Population growth and Earth's human carrying capacity. *Science* 269:341–346.

Conway, K. 1993. State expenditures on federally listed threatened and endangered species. *Endangered Species Update* 10:5–8.

Conway, W. G. 1986. The practical difficulties and financial implications of endangered species breeding programmes. *Int. Zoo Yb.* 24/25:210–219.

Costanza, R. 1991. *Ecological Economics: The Science and Management of Sustainability.* Columbia University Press, New York.

Currie, D. J., and V. Paquin. 1987. Large-scale biogeographical patterns of species richness of trees. *Nature* 329:326–327.

Daily, G. C., and P. R. Ehrlich. 1992. Population, sustainability, and Earth's carrying capacity. *Bioscience* 42:761–771.

Dawson, W. R., et al. 1987. Report of the scientific advisory panel of the spotted owl. *Condor* 89:205–229.

Diamond, J. 1987. Extant unless proven extinct? Or, extinct until proven extant? *Con. Biol.* 1:77–79.

———. 1995. Easter's End. *Discover* 16(6):62–69.

Diamond, J. M. 1975. The island dilemma: Lessons of modern biogeographic studies for the design of natural reserves. *Biol. Conserv.* 7:129–146.

Dobson, A. P., and E. R. Carper. 1993. Health and Climate Change: Biodiversity. Lancet 342:1096–1099.

Dobson, A. P., and A. M. Lyles. 1989. The population dynamics and conservation of primate populations. *Con. Biol.* 3:362–380.

Durning, A. T. 1992. Guardians of the land: Indigenous peoples and the health of the Earth. *Worldwatch Papers* 112:1–62.

———. 1993. Saving the forests: What will it take? *Worldwatch Papers* 117:1–51.

Eisner, T., J. Lubchenco, E. O. Wilson, D. S. Wilcove, and M. J. Bean. 1995. Building a scientifically sound policy for protecting endangered species. *Science* 268:1231–1232.

Elliot, M. 1991. A survey of world travel and tourism. *Economist* March 23, 1–22.

Elton, C. S., and M. Nicholson, 1942. The ten-year cycle in numbers of the lynx in Canada. *J. Anim. Ecol.* 11:215–244.

Erwin, T. L. 1991. An evolutionary basis for conservation management. *Science* 253:750–752.

Fenchel, T. 1993. There are more small than large species. *Oikos* 68:375–378.

Fitzgerald, S. 1989. *International Wildlife Trade: Whose Business is it?* World Wildlife Fund, Washington, D.C.

Foose, T. J. 1993. Riders of the last ark: The role of captive breeding in conservation strategies. In L. Kaufman and K. Mallory, eds. *The Last Extinction,* 149–178. MIT Press, Cambridge, Mass.

French, H. F. 1993. Costly tradeoffs: Reconciling trade and the environment. *Worldwatch Papers* 113:1–74.

Gaston, K. J. 1994. *Rarity.* Chapman & Hall, London.

Gaston, K. J., and R. M. May. 1992. Taxonomy of taxonomists. *Nature* 356:281–282.

Gilpin, M., and I. Hanski. 1991. *Metapopulation Dynamics: Empirical and Theoretical Investigations.* Academic Press, London.

Gipps, J. H. W. 1991. *Beyond Captive Breeding: Re-introducing Endangered Mammals of the World.* Oxford Science Publications, Oxford.

Glick, D., M. Carr, and B. Harting. 1991. *An Environmental Profile of the Greater Yellowstone Ecosystem.* Greater Yellowstone Coalition, Bozeman, Montana.

Goude, A. 1990. *The Human Impact on the Natural Environment.* 3d ed. MIT Press, Cambridge, Mass.

Grassle, J. F. 1991. Deep-sea benthic biodiversity. *Bioscience* 41:464–469.

Graveland, J., R. van der Wal, J. H. van Balen, and A. J. van Noordwijk. 1994. Poor reproduction in forest passerines from decline of snail abundance on acidified soils. *Nature* 368:446–448.

Griffith, B., J. M. Scott, J. W. Carpenter, and C. Reed. 1989. Translocation as a species conservation tool: Status and strategy. *Science* 245:477–480.

Hardin, G. 1968. The tragedy of the commons. *Science* 162:1243–1248.

———. 1993. *Living Within Limits: Ecology, Economics, and Population Taboos.* Oxford University Press, Oxford.

Harrison, S. 1994. Metapopulations and conservation. In P. J. Edwards, R. M. May, and N. R. Webb, eds. *Large-scale Ecology and Conservation Biology,* 111–128. Blackwell Scientific, Oxford.

Hedrick, P. W., and P. S. Miller. 1992. Conservation genetics: Techniques and fundamentals. *Ecol. Appl.* 2:30–46.

Hobbs, R. J. 1992. The role of corridors in conservation—solution or bandwagon. *Trends in Ecology & Evolution* 7:389–392.

Hodell, D. A., J. H. Curtis, and M. Brenner. 1995. Possible role of climate in the collapse of Classic Maya civilization. *Nature* 375:391–394.

Horn, H. H. 1993. Biodiversity in the backyard. *Sci. Am.* 268 (Jan.): 150–152.

Horwich, R. H., F. Koontz, E. Saqui, H. Saqui, and K. Glander. 1993. A reintroduction program for the conservation of the black howler monkey in Belize. *Endangered Species Update* 10:1–6.

Huston, M. A. 1994. *Biological Diversity: The Coexistence of Species on Changing Landscapes.* Cambridge University Press, Cambridge.

Janzen, D. H. 1986. The external threat. In M. E. Soule, ed. *Conservation Biology. The Science of Scarcity and Diversity,* 286–303. Sinauer, Northampton, Mass.

Kareiva, P. M., J. G. Kingsolver, and R. B. Huey. 1993. *Biotic Interactions and Global Change.* Sinauer, Sunderland, Mass.

Keiter, R. B., and M. S. Boyce. 1991. *The Greater Yellowstone Ecosystem: Redefining America's Wilderness Heritage.* Yale University Press, New Haven.

Kellert, S. R. 1985. Social and perceptual factors in endangered species management. *J. Wildl. Manage.* 49:528–536.

Kinzig, A. P., and R. H. Socolow. 1994. Human impacts on the nitrogen cycle. *Phys. Today* 24–31.

Kleiman, D. G. 1989. Reintroduction of captive mammals for conservation. *Bioscience* 39:152–161.

Lande, R. 1988. Genetics and demography in biological conservation. *Science* 241:1455–1460.

———. 1993. Risks of population extinction from demographic and environmental stochasticity and random catastrophes. *Am. Nat.* 142:911–927.

Lande, R., S. Engen, and B-E. Saether. 1994. Optimal harvesting, economic discounting and extinction risk in fluctuating populations. *Nature* 372:88–90.

Lawton, J. H. 1990. Species richness and population dynamics of animal assemblages. Patterns in body size: Abundance space. *Phil. Trans. R. Soc. Lond. B* 330:283–291.

Leader-Williams. N., S. D. Albon, and P. S. M. Berry. 1990. Illegal exploitation of black rhinoceros and elephant populations: Patterns of decline, law enforcement and patrol effort in Luangwa Valley, Zambia. *J. Appl. Ecol.* 27:1055–1087.

Leigh Jr., E. G., S. J. Wright, E. A. Herre, and F. E. Putz. 1993. The decline of tree diversity on newly isolated tropical islands: A test of a null hypothesis and some implications. *Evol. Ecol.* 7:76–102.

Losos, E., J. Hayes, A. Phillips, D. Wilcove, and C. Alkire. 1995. Taxpayer-subsidized resource extraction harms species. *Bioscience* 45: 446–455.

Lubchenco, J., et al. 1991. The sustainable biosphere initiative: An ecological research agenda. *Ecology* 72:371–412.

Ludwig, D., R. Hilborn, and C. Walters. 1993. Uncertainty, resource exploitation, and conservation: Lessons from history. *Science* 260:17–36.

Lyles, A. M., and R. M. May. 1987. Problems in leaving the ark. *Nature* 326:245–246.

MacArthur, R. H. 1972. *Geographical Ecology: Patterns in the Distribution of Species.* Harper & Row, New York.

MacArthur, R. H., and E. O. Wilson. 1967. *The Theory of Island Biogeography.* Princeton University Press, Princeton.

Mace, G. M., and R. Lande. 1991. Assessing extinction threats: Toward a reevaluation of IUCN threatened species categories. *Con. Biol.* 5:148–157.

Mann, C. C., and M. L. Plummer. 1995. *Noah's Choice: the Future of Endangered Species.* Alfred A. Knopf, New York.

Martin, E. B., and L. Vigne. 1989. The decline and fall of India's ivory industry. *Pachyderm* 12:4–21.

May, R. M. 1973. *Stability and complexity in model ecosystems.* Princeton University Press, Princeton.

———. 1988. How many species are there on earth? *Science* 241:1441.

———. 1990b. How many species? *Phil. Trans. R. Soc. Lond. B* 330: 293–304.

May, R. M., J. H. Lawton, and N. E. Stork. 1995. Assessing extinction rates. In J. H. Lawton and R. M. May, eds. *Extinction Rates.* Oxford University Press, Oxford.

May, R. M., and A. M. Lyles. 1987. Living latin binomials. *Nature* 326:642–643.

McMahan, L. R. 1990. Propagation and reintroduction of imperiled plants, and the role of botanical gardens and arboreta. *Endangered Species Update* 8:4–7.

McNaughton, S. J. 1978. Stability and diversity of ecological communities. *Nature* 274:251–252.

———. 1985. Ecology of a grazing ecosystem: The Serengeti. *Ecol. Monogr.* 55:259–294.

Meagher, M. 1989. Range expansion by bison of Yellowstone National Park. *J. Mamm.* 70:670–675.

Meffe, G. K., and C. R. Carroll. 1994. *Principles of Conservation Biology.* Sinauer, Sunderland, Mass.

Menges, E. S. 1990. Population viability analysis for an endangered plant. *Con. Biol.* 4:52–62.

Meyer, A. 1993. Phylogenetic relationships and evolutionary processes in East African cichlid fishes. *TREE* 8:279–284.

Miller, R. R., J. D. Williams, and J. E. Williams. 1989. Extinctions of North American fishes during the past century. *Fisheries* 14: 22–38.

Mills, L. S., M. E. Soule, and D. F. Doak. 1993. The keystone-species concept in ecology and conservation. *Bioscience* 43:219–224.

Milner-Gulland, E. J., and J. R. Beddington. 1993. The exploitation of elephants for the ivory trade: An historical perspective. *Proc. R. Soc. Lond. B* 252:29–37.

Mittermeier. R. A. 1986. Primate diversity and the tropical forest. In E. O. Wilson, ed. *Biodiversity.* National Academy of Sciences, Washington, D.C.

Nee, S., P. H. Harvey, and R. M. May. 1991a. Lifting the veil on abundance patterns. *Proc. R. Soc. Lond. B* 243:161–163.

New, T. R. 1991. *Butterfly Conservation.* Oxford University Press, Oxford.

Newmark, W. D. 1986. Species-area relationship and its determinants for mammals in western North American national parks. *Biol. Jnl. Linn. Soc.* 28:83–98.

Norton, B. G. 1986. *The Preservation of Species.* Princeton University Press, Princeton.

Oldfield, M. L. 1984. *The Value of Conserving Genetic Resources.* Sinauer, Sunderland, Mass.

Padoch, C, and A. Susanto. 1993. Fruits of diversity. Pacific Discovery 46:30–35.

Parsons, P. A. 1990. The metabolic cost of multiple environmental stresses—implications for climatic change and conservation. *Trends in Ecology & Evolution* 5:315–317.

Pauly, D., and V. Christensen. 1995. Primary production required to sustain global fisheries. *Nature* 374:255–257.

Peters, C. M., A. H. Gentry, and R. O. Mendelsohn. 1989. Valuation of an Amazonian rainforest. *Nature* 339:655–656.

Peters, R. L., and T. E. Lovejoy. 1992. *Global Warming and Biological Diversity.* Yale University Press, New Haven.

Pimentel, D., et al. 1992. Environmental and economic costs of pesticide use. *Bioscience* 42:750–760.

Pimentel, D., and L. Levitan. 1986. Pesticides: Amounts applied and amounts reaching pests. *Bioscience* 36:86–91.

Pimm, S. L., J. H. Lawton, and J. E. Cohen. 1991. Food web patterns and their consequences. *Nature* 350:669–674.

Pimm, S. L., G. J. Russell, J. L. Gittleman, and T. M. Brooks. 1995. The future of biodiversity. *Science* 269:347–350.

Ponting, C. 1991. *A Green History of the World.* Penguin, London.

Poole, J. H., and A. P. Dobson. 1992. Why the ban must stay. *Con. Biol.*

———. 1996. Ivory poaching and the viability of African elephant populations. *Con. Biol.* (in press)

Poole, J. H., and J. B. Thomsen. 1989. Elephants are not beetles: Implications of the ivory trade for the survival of the African elephant. *Oryx* 23:188–198.

Postel, S. 1995. Where have all the rivers gone? *World Watch* 8:9–19.

Prendergast, J. R., R. M. Quinn, J. H. Lawton, B. C. Eversham, and D. W. Gibbons. 1993. Rare species, the coincidence of diversity hotspots and conservation strategies. *Nature* 365:335–337.

Pressey, R. L., C. J. Humphries, C. R. Margules, R. I. Vane-Wright, and P. H. Williams. 1993. Beyond opportunism: Key principles for systematic reserve selection. *TREE* 8:124–128.

Prins, H. H. T., and H. P. van der Jeugd. 1993. Herbivore population clashes and woodland structure in East Africa. *Journal of Ecology* 8:305–314.

Rabinowitz, D., S. Cairns, and T. Dillon. 1986. Seven forms of rarity and their frequency in the flora of the British Isles. In M. E. Soule, ed. *Conservation Biology: The Science of Scarcity and Diversity,* 182–204. Sinauer, Sunderland, Mass.

Rasker, R., N. Tirell, and D. Kloepfer. 1992. *The Wealth of Nature: New Economic Realities in the Yellowstone Region.* Wilderness Society, Washington, D.C.

Raup, D. M. 1986. Biological extinction in Earth history. *Science* 231:1528–1533.

Redford, K. H. 1992. The empty forest. *Bioscience* 42:412–422.

Reed, J. M. 1992. A system for ranking conservation priorities for neotropical migrant birds based on relative susceptibility to extinction. In J. M. Hagan III and D. W. Johnston, eds. *The Ecology and Conservation of Neotropical Migrant Landbirds.* Smithsonian Institution Press, Washington, D.C.

Reed, J. M., and A. P. Dobson. 1993. Behavioural constraints and conservation biology: Conspecific attraction and recruitment. *TREE* 8:253–255.

Reid, W. V., 1992. How many species will there be? In T. C. Whitmore, and J. A. Sayer, eds. *Tropical Deforestation and Species Extinction,* 55–73. Chapman & Hall/IUCN, London.

Reid, W. V., A. Sittenfeld, S. A. Laird, D. H. Janzen, C. A. Meyer, M. A. Gollin, R. Gamez, and J. Calestous. 1993. *Biodiversity Prospecting: Using Genetic Resources for Sustainable Development.* World Resources Institute, Washington.

Richter-Dyn, N., and N. S. Goel. 1972. On the extinction of colonizing species. *Theor. Pop. Biol* 3:406–433.

Robinson, J. G., and K. H. Redford. 1991. *Neotropical Wildlife Use and Conservation.* University of Chicago Press, Chicago.

Robinson, S. K., F. R. Thompson III, T. M. Donovan, D. R. Whitehead, and J. Faaborg. 1995. Regional forest fragmentation and the nesting success of migratory birds. *Science* 267:1987–1990.

Schneider, S. H. 1989. The greenhouse effect: Science and policy. *Science* 243:771–782.

Scott, J. M., B. Csuti, J. D. Jacobi, and J. E. Estes. 1987. Species richness. *Bioscience* 37:782–788.

Scott, J. M., et al. 1993. Gap analysis: A geographic approach to protection of biological diversity. *Wildl. Monogr.* 123:1–41.

Shaffer, M. L. 1981. Minimum population sizes for species conservation. *Bioscience* 31:131–134.

Signor, P. W. 1990. The geologic history of diversity. *Annu. Rev. Ecol. Syst.* 21:509–539.

Silvertown, J., M. E. Dodd, K, McConway, J. Potts, and M. J. Crawley. 1994. Rainfall, biomass variation, and community composition in the Park Grass Experiment. *Ecology* 75:2430–2437.

Simenstad, C. A., J. A. Estes, and K. W. Kenyon. 1978. Aleuts, sea-otters, and alternate stable-state communities. *Science* 200:403–409.

Sinclair, A. R. E., and P. Arcese. 1995. *Serengeti II: Dynamics, Management, and Conservation of an Ecosystem*. Chicago University Press, Chicago.

Sinclair, A. R. E., and J. M. Fryxell. 1985. The Sahel of Africa: Ecology of a disaster. *Can. J. Zool.* 63:987–994.

Sinclair, A. R. E., and M. Norton-Griffiths. 1979. *Serengeti: Dynamics of an Ecosystem*. Chicago University Press, Chicago.

Skole, D. L., W. H. Chomentowski, W. A. Salas, and A. D. Nobre. 1994. Physical and human dimensions of deforestation in Amazonia. *Bioscience* 44:314–322.

Snyder, N. F. R., and H. A. Snyder. 1989. Biology and conservation of the California Condor. *Current Ornithology* 6:175–267.

Steadman, D. W. 1989. Extinction of birds in Eastern Polynesia: A review of the record, and compari-sons with other Pacific groups. *Jnl. Arch. Sci.* 16:177–205.

Stevenson, M., T. J. Foose, and A. Baker. 1991. *Global Captive Action Plan for Primates*. Conservation International, Washington.

Templeton, A. R. 1990. The role of genetics in captive breeding and reintroduction for species conservation. *Endangered Species Update* 8:14–17.

Terborgh, J. 1974. Preservation of natural diversity: The problem of extinction prone species. *Bioscience* 24:715–722.

———. 1989. *Where Have All the Birds Gone?* Princeton University Press, Princeton.

Terborgh, J., and B. Winter. 1980. Some causes of extinction. In M. S. Soule and B. A. Wilcox, eds. *Conservation Biology. An Evolutionary-Ecological Perspective*, 119–133. Sinauer, Sunderland, Mass.

Tilman, D., and J. A. Downing. 1994. Biodiversity and stability in grasslands. *Nature* 367:363–365.

Tilman, D., R. M. May, C. L. Lehman, and M. A. Nowak. 1994. Habitat destruction and the extinction debt. *Nature* 371:65–66.

Tudge, C. 1992. *Last Animals at the Zoo*. Island Press, Washington.

Tuomisto, H., K. Ruokolainen, R. Kalliola, A. Linna, W. Danjoy, and Z. Rodriguez. 1995. Dissecting Amazonian biodiversity. *Science* 269:63–66.

Turner II, B. L., W. C. Clark, R. W. Kates, J. F. Richards, J. T. Mathews, and W. B. Meyer. 1990. *The Earth as Transformed by Human Action*. Cambridge University Press, Cambridge.

Vane-Wright, R. I., C. J. Humphries, and P. H. Williams. 1991. What to protect? Systematics and the agony of choice. *Biol. Conserv.* 55:235–254.

Vitousek, P. M. 1994. Beyond global warming: Ecology and global change. *Ecology* 75:1861–1876.

Vitousek, P. M., and L. R. Walker. 1989. Biological invasion by Myrica faya in Hawai'i: Plant demography, nitrogen fixation, ecosystem effects. Ecol. Monogr. 59:247–265.

Weber, P. 1993. Abandoned seas: Reversing the decline of the oceans. *Worldwatch Papers* 116:1–66.

Wells, M., K. Brandon, and L. Hannah. 1992. *People and Parks: Linking Protected Area Management with Local Communities*. World Bank, World Wildlife Fund, US-AID, Washington, D.C.

Wilcove, D. S. 1994. Turning conservation goals into tangible results: The case of the spotted owl and old-growth forests. In P. J. Edwards, R. M. May, and N. R. Webb, eds. *Large-Scale Ecology and Conservation Biology*, 313–329. Blackwell Scientific, Oxford.

Wilcove, D. S., C. H. McLellan, and A. P. Dobson. 1986. Habitat fragmentation in the temperate zone. In M. E. Soule, ed. *Conservation Biology: The Science of Scarcity and Diversity*, 237–256. Sinauer, Sunderland, Mass.

Wilson, E. O. 1988. *Biodiversity*. National Academy Press, Washington.

Wright, S. J., and S. P. Hubbell. 1983. Stochastic extinction and reserve size: a focal species approach. *Oikos* 41:466–476.

Sources of Illustrations

Prologue   *p. 2:* James Martin/Tony Stone Images. *p. 4:* Will Landon/Tony Stone Images. **Chapter 1**   *p. 6:* Art Wolfe. *p. 8:* Art Wolfe. *p. 9:* From I. Tattersall, *The Primates of Madagascar*, Columbia University Press, 1982, figs. 3.3 and 3.26. *p. 10:* Bob Abraham/Pacific Stock. *p. 11:* Chip Clark. *pp. 12 and 13:* C. B. Williams. *Patterns in the Balance of Nature and Related Problems in Quantitative Ecology*, Academic Press, London, 1964. *p. 14:* From S. Nee et al., *Proc. R. Soc. Lond. B* 243:161–163, 1991, fig. 4. *p. 16:* From S. Nee et al., *Nature* 351:312–313 (May 23, 1991). *p. 20:* From Robert H. MacArthur, *Geographical Ecology*, Harper and Row, 1972, fig. 8.8. *p. 21:* From D. J.

Currie and V. Paquin, *Nature* 329:326–327, 1987, fig. 1. *p. 23:* From Robert M. May, *Science* 241:1441, 1988, fig. 6. *p. 24:* Tim Flannery, Australian Museum. *pp. 25 and 27:* From Robert M. May, *Scientific American* 267(4), October 1992. *p. 26:* Linda L. Sims. *p. 29:* Kari Rene Hall, *Los Angeles Times*. **Chapter 2**   *p. 32:* Milan Horacek/Bilderberg. *p. 34:* Explorer Archives. *p. 35:* Michael Nichols/Magnum. *p. 36:* From D. S. Wilcove et al., in M. E. Soule (ed.), *Conservation Biology*, Sinauer Associates, 1986, fig. 1. *p. 38:* From M. Williams, *Americans and their Forests*, Cambridge University Press, 1989. *p. 41:* R. A. Houghton, *Bioscience* 44:305–313, 1994. *p. 44:*

Betty Press/Woodfin Camp and Associates. *p. 47*: R. H. MacArthur and E. O. Wilson, *The Theory of Island Biogeography*. Princeton University Press, Princeton, 1967. T. G. Shreever and C. F. Mason, *Oecologia* 45:414–418, 1980. *p. 49*: Jeff Foott/Bruce Coleman. *p. 50*: Terborgh, *Where have all the birds gone?* Princeton University Press, Princeton, 1989. *p. 51*: From Robert Askins, *Science* 267:1957, 1995. *p. 52*: From S. K. Robinson et al., *Science* 267:1997–1980, 1995. *p. 55*: H. Andren, *Oikos* 71:355–366, 1995. *p. 57*: Martin Wendler/Natural History Photo Agency. **Chapter 3** *p. 58*: Christopher Fransella/ Raymond Fortt Studios/National Museums of Scotland. *p. 60*: Andrew Dobson. *p. 61*: M. V. Lomolino and R. Channell, *J. Mammal.* 76:335–347, 1995. p. 63: Rosamond Purcell, Bishop Museum *p. 64*: Stephen Krasemann/Natural History Photo Agency. *p. 65*: Bishop Museum. *p. 66*: W. V. Reid and K. R. Miller, *Keeping Options Alive: The Scientific Basis for Conserving Biodiversity*, World Resources Institute, 1987. *p. 72*: LuRay Parker, Wyoming Game and Fish Department. *p. 73*: S. C. Forrest et al., *J. Mammal.* 69:261–273, 1988. *p. 76*: Art Wolfe. *p. 77*: R. H. MacArthur, *Geographical Ecology: Patterns in the Distribution of Species*, Harper & Row, New York, 1972. *p. 80*: Peggy and Erwin Bauer. *p. 79*: S. L. Pimm, H. L. Jones, and J. Diamond, *Am. Nat.* 132:757–185, 1988. *p. 83*: Robert Rattner. **Chapter 4** *p. 86*: Richard T. Bryant. *p. 90*: Greg Vaughn/Tony Stone Images. *pp. 91, 92*: From D. Wilcove, in P. J. Edwards, R. M. May, and N. R. Webb (eds.), *Large-Scale Ecology and Conservation Biology*, Blackwell Scientific, 1994, fig. 14. *p. 93*: M. Burgman, S. Ferson, and H. R. Akcakaya, *Biol. Conserv.* 43:925, 1988. *p. 95*: Michael Nichols/Magnum. *p. 104*: Don Madden/Hedgehog House. *pp. 97 and 99*: F. Campbell, *Endangered Species Update* 5:20–26, 1988. *p. 107*: From R. I. Vane Wright et al., *Biol. Conserv*, 55: 235-254, 1991. *p. 108*: Axel Mayer, State University of New York at Stony Brook. **Chapter 5** *p. 110*: Ted Kerasote/Photo Researchers. *p. 112*: The Whaling Museum, New Bedford, MA. *p. 113*: From Ray Hilborn, in ClarkTurner, et al. (eds.), *The Earth as Transformed by Human Action*, Cambridge University Press, 1993. *p. 114*: Colin Monteath/Hedgehog House. *p. 116*: J. Cherfas, *New Scientist* 5 June:36–40, 1986. *p. 118*: Terrance Moore. *p. 121*: Klaus D. Francke/Bilderberg. *p. 123*: P. Weber, *Worldwatch Papers* 120:1–76, 1994. *p. 124*: Pat Hathaway Historical Collection. *p. 128*: Stoddart/Katz Pix/Woodfin Camp. *p. 129*: Stoddart/Katz Pix/Woodfin Camp. *p. 134*: Steve Robinson/Natural History Photo Agency. **Chapter 6** *p. 136*: Andrew Dobson. *p. 138*: Wildlife Conservation Society/New York Zoological Society. *p. 139*: Michael Nichols/Magnum. *p. 141*: From N. F. R. Snyder and H. A. Snyder, *Current Ornithology* 6, 1989, fig. 1. *p. 143*: N. Takahata, *Science* 267:35 (6 January, 1995). *p. 146*: Alberto Nardi/Natural History Photo Agency. *p. 147*: From *Endangered Species Bulletin*. *p. 150*: Andrew Dobson. *p. 152*: *International Zoo Yearbook*. *p. 152*: (left) Ron Garrison, Zoological Society of San Diego; (right) Ken Kelley, Zoological Society of San Diego. *p. 154*: Daniel Heuclin/Natural History Photo Agency. *p. 157*: Lynn Johnson/Black Star. *p. 159*: G. Ziesler, Peter Arnold, Inc. **Chapter 7** *p. 162*: Yellowstone National Park Archives. *p. 167*: W. D. Newmark, *Nature* 325:430–432, 1987. *p. 168*: B. Groomsbridge, *Global Biodiversity*, World Conservation Monitoring Centre, Chapman Hall, 1992. *p. 169*: From J. Diamond, *Biol. Conserv.* 7:129–146, 1975, fig. 7. *p. 171*: From Myers, in M. E. Soule (ed.), *Conservation Biology*, Sinauer Associates, 1986, fig. 1. *p. 172*: From C. J. Bibby et al., *Putting Biodiversity on the Map*, International Council for Bird Preservation, 1992, fig. 28. *p. 173*: D. L. Pearson and F. Cassola, *Con. Biol.* 6:376–391, 1992. *p. 174*: From J. M. Scott et al., *BioScience* 37:782–788, 1987, figs. 2 and 3. *p. 175*: Jack Jeffrey. *p. 178*: Jim Peaco, National Park Service, Yellowstone. **Chapter 8** *p. 180*: Norbert Wu. *p. 182*: Kenneth J. Howard/Bonnie Kamin. *pp. 184 and 186*: From J. A. Estes and J. F. Palmisano, *Science* 185:1058–1060, 1974, figs. 1 and 3. *p. 187*: H. Rezit Akcakaya, *Ecol. Monographs* 62(1):134. *p. 190*: R. L. Matthews/Planet Earth Pictures. *p. 192*: Andrew Dobson. *p. 194*: M. J. Crawley, *Plants Today* 1:152–158. *p. 196*: Erwin and Peggy Bauer. *p. 197*: Tim Crawford. *p. 200*: (top left) G. Bagley; (top and bottom right) M. Meagher. *p. 201*: From A. D. Bradshaw, *Proc. R. Soc. Lond.* B 223:1–23, 1984, fig. 3. *p. 203*: Juliette Murguia. *p. 206*: From K. Campbell and H. Hofer, in A. R. E. Sinclair and P. Arcese (eds.), *Serengeti II*, University of Chicago Press, 1995, fig. 25.5. *p. 207*: William Conway. **Chapter 9** *p. 210*: Betty Press/Woodfin Camp. *p. 213*: J. Adams, *WWF & CF Letter* 2:1–8, 1990. *p. 214*: Andrew Dobson. *p. 216*: Table from P. M. Vitousek et al., *Bioscience* 36:368–373, 1986. *p. 219*: I. Newton and I Wyllie, *J. Appl. Ecol.* 29:476–484, 1993. *p. 222*: P. M. Vitousek, *Ecology* 75:1861–1876, 1994. *p. 223*: From Margaret Davis, in R. L. Peters and T. E. Lovejoy (eds.), *Global Warming and Biological Diversity*, Yale University Press, 1992, fig. 22.3B. *p. 224*: From Murphy and Weiss, in R. L. Peters and T. E. Lovejoy (eds.), *Global Warming and Biological Diversity*, Yale University Press, 1992, fig. 26.3. *p. 226*: Michael Nichols/Magnum. **Chapter 10** *p. 230*: Joel Rogers/Tony Stone Images. *p. 232*: Andrew Dobson. *p. 234*: Anthony Bannister/Natural History Photo Agency. *p. 236*: Art Wolfe. *p. 240*: Christine Padoch, New York Botanical Garden. *p. 242*: Michael J. Balick/Peter Arnold, Inc. *p. 243*: Biophoto Associates/Photo Researchers. *p. 246*: Michael Nichols/Magnum. *p. 252*: Planet Earth Pictures.

Index